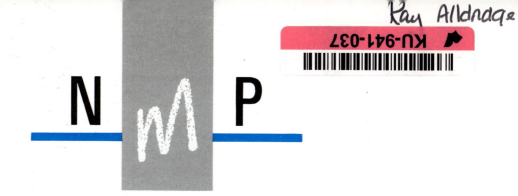

N M P

This book was written by

Eon Harper
Dietmar Küchemann
Michael Mahoney
Sally Marshall
Edward Martin
Heather McLeay
Peter Reed
Sheila Russell

NMP Director
Eon Harper

NMP Research
Edward Martin

MATHEMATICS FOR **5** SECONDARY SCHOOLS

BLUE TRACK

Longman

The front cover shows a detail from *Abstract Oil Composition*, by
Edward Wolfe (the painting is reproduced in full on the back cover).
Edward Wolfe was born in Johannesburg in 1897. He was a painter
of figures, floras and landscapes and was influenced by Cezanne and
Matisse.

Edward Wolfe *Abstract Oil Composition*
by courtesy of James O'Connor
(photo Bridgeman Art Library)

Longman Group UK Limited
Longman House, Burnt Mill, Harlow, Essex CM20 2JE, England
and Associated Companies throughout the world.

First published 1990
ISBN 0 582 22519 1

Set in 11/13 Times

Printed in Great Britain by Scotprint Limited, Musselburgh

PREFACE

This book and the three companion books, Year 4 Blue Track, Year 4 Red Track and Year 5 Red Track cover all Levels in Key Stage 4 of the National Curriculum and provide a complete two-year Examination course for all grades of the GCSE Examination.

The course builds upon the Years 1–3 Red Track and Blue Track course of *NMP Mathematics for Secondary Schools*. Each book contains Review sections which revise the mathematics introduced in Years 1–3. The course can thus be used by schools which have hitherto not used NMP materials in the foundation years.

The Blue Track books for Years 4 and 5 provide for GCSE Grades C/D to G; the Red Track books provide for Grades A to C/D. Further details can be found in the Teachers' File for each track.

NMP was founded in 1981 to develop teaching and learning materials suited to the emerging new syllabuses and National Criteria. The material was researched and evaluated at the University of Bath and written by practising teachers and professional educators.

Each text provides for pupil–pupil and pupil–teacher discussion, oral and mental work, skill and practice work, written and calculator work, problem solving, investigation and extended assignments for GCSE coursework. Each of these aspects is integrated into the text to provide the variety of learning opportunities required by the various GCSE Boards and the National Curriculum.

The text falls into two Sections, A and B, whose Chapters contain Review, Consolidation, Core and Enrichment material. The relationship of these to GCSE needs is explained in detail in the Teachers' File.

ACKNOWLEDGEMENTS

We are grateful to the following for permission to reproduce photographs:
Heather Angel, pages 87 *below*, 90; Animal Photography, pages 148, 149 (photos Sally Anne Thompson); Beken of Cowes, page 95; John Birdsall, pages 40 *above*, 101; Black & Decker, page 6 *above left*; Camera Press, page 2 (photo Jon Blau); Canon (UK), page 14 *right*; J. Allan Cash, page 34; Bruce Coleman, page 138; The Solomon R. Guggenheim Museum, New York, page 118 (photo Robert E. Mates); Honda (UK), page 63; Hoover plc, page 119; Ministry of Defence, page 83 *left*; Network, page 27 (photo Laurie Sparham); Philips, pages 1, 62; Photo Co-op, pages 32 (photo Crispin Hughes), 57 (photo Anna Arnone); Professor John Postgate FRS, page 91 *below*; Rex Features, page 185; Science Photo Library, pages 40 *below* (photo NASA), 42 (photo ESA), 65 (photo Adam Hart-Davis), 87 *above* (photo CNRI), 91 *above* (photo Dr. Jeremy Burgess), 91 *centre left* (photo London School of Hygiene & Tropical Medicine), 91 *centre right* (photo Barry Dowsett), 137 (photo NASA); Harry Smith Horticultural Photographic Collection, page 121 *left*; Michael Warren, page 121 *right*; Janine Wiedel Photo Library, page 68.

Other copyright material:
British Telecommunications plc, page 53; Department of Health and Social Security, page 28, extract from leaflet FB23 '*Young people's guide to Social Security*'. Richard Rogers Partnership Ltd, page 84.

Illustrations by Allan Lamb and Oxford Illustrators, cartoons by Martin Shovel.

CONTENTS

REVIEW

- 15% of £300 means $\frac{15}{100}$ of £300, or $0.15 \times £300$.

To find 15% of £300 on a calculator we press:

$\boxed{\text{C}}\ \boxed{3}\ \boxed{0}\ \boxed{0}\ \boxed{\times}\ \boxed{1}\ \boxed{5}\ \boxed{\div}\ \boxed{1}\ \boxed{0}\ \boxed{0}\ \boxed{=}$

or $\boxed{\text{C}}\ \boxed{0}\ \boxed{.}\ \boxed{1}\ \boxed{5}\ \boxed{\times}\ \boxed{3}\ \boxed{0}\ \boxed{0}\ \boxed{=}$

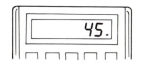

You would save £45 on this £300 midi system.

- You save £5 on the £50 radio. The percentage reduction is $\frac{5}{50} = \frac{10}{100} = 10\%$

■ What percentage of the original price is the sale price?

- We can estimate percentage calculations like this:

          48% of 197 m

      ↗   ↖

about 50%   about 200 m

50% of 200 m is 100 m. So 48% of 197 m is about 100 m.

■ Without using a calculator, explain how you know that 23% of £37.50 is about £9.

- You should know these equivalences:

1%	=	$\frac{1}{100}$			=	0.01
5%	=	$\frac{5}{100}$	=	$\frac{1}{20}$	=	0.05
10%	=	$\frac{10}{100}$	=	$\frac{1}{10}$	=	0.1
20%	=	$\frac{20}{100}$	=	$\frac{1}{5}$	=	0.2
25%	=	$\frac{25}{100}$	=	$\frac{1}{4}$	=	0.25
50%	=	$\frac{50}{100}$	=	$\frac{1}{2}$	=	0.5
75%	=	$\frac{75}{100}$	=	$\frac{3}{4}$	=	0.75
33%	=	$\frac{33}{100}$	$\approx$	$\frac{1}{3}$	$\approx$	0.33

A1

CONSOLIDATION

A1

Do not use a calculator. Write down only the result. (This does not mean you will be able to do each question very quickly. Questions 3, 4 and 5 will require careful thinking.)

1   Write each of these as a decimal calculation (for example, 24% of 8 kg = 0.24 × 8 kg). You do not have to do the calculation.

   a) 33% of 7 m   b) 5% of 17 kg.

2   Write each of these as a percentage calculation (for example, $\frac{3}{10}$ of £5.60 = 30% of £5.60):

   a) $\frac{7}{10}$ of £2.30   b) $\frac{3}{4}$ of 2.8 tonnes   c) $\frac{1}{5}$ of 8.6 kg.

3   Estimate each of these:

   a) 33% of £12   b) 11% of £90   c) 47% of 18.1 kg   d) 73% of 29 ml.

4   A watch is reduced from £3.99 to £3 in a sale. Roughly, what is the percentage reduction?

5   In a sale all electrical items are reduced by 12%. The original price of a refrigerator is £298. Roughly, what is the sale price?

6   Use a calculator to do the calculations in questions 1 and 2.

CORE

Buying and selling

1   Shopkeepers buy their goods from wholesalers. (A wholesaler sells articles in large quantities to the shopkeeper ('retailer') who then sells them on to the customers.)
   The price a shopkeeper pays the wholesaler is called the *cost price*.

To calculate the price to charge the customer (called the *selling price*) the shopkeeper:

- adds a percentage amount (say 50%) to the price paid to the wholesaler. This is to cover 'overhead' costs such as transporting the goods, heating and lighting the shop, paying wages to assistants, etc. It is called the '*mark-up*' percentage.
- may have to add a percentage required by law for VAT (Value Added Tax).
  (Some goods are exempt from VAT, such as (in 1989) children's clothes, and most food.)

a) A shopkeeper buys training shoes from a wholesaler for £8, and adds a 'mark up' of 60%. Check that the price is now £12.80.

b) VAT at 15% is added to the £12.80. Check that the selling price is £14.72.

c) How much would you pay for a pair of shoes from the same shop for which the shopkeeper paid £20?

CHALLENGE

2 This is a bill from a builder's merchant.

a) What percentage VAT do you think has been added? Explain how you arrived at your result.

b) The builder's merchant adds a 60% mark up to his own cost prices.

  (i) Check that the cost price of the plastic bucket was 50p.

  (ii) Roughly, how much did the builder's merchant pay for each bag of sand? (That is, what is the cost price of a bag of sand?)

CASH SALE -COLLECTION
WELMSLOW BUILDERS MERCHANTS
BUILDERS & PLUMBERS MERCHANTS, TOOL HIRE, SALES & REPAIRS

PARK ROAD, WELMSLOW, STRATFORD SD 16 4JF. TEL (90419 564411)

VAT Reg. No 301 418511

NAME: **BMF**
ADDRESS: **MEMBER**

DATE 15.8.90 Cash | Cheque | Sold by | Cust. Order No.

Quantity	Description	Unit Price	Zero VAT	Std VAT	
1	Plastic Bucket.			80	
1	Bag of Sand.		1	00	
		Column Totals			
		Net Goods	1	80	
		VAT		27	
		TOTAL	2	07	

3 The U-frame Tyre Centre advertises tyres for these prices:

Calculate the total for the order shown on the bill, if VAT is 15%.

size 145 x 13
£17 +VAT

size 165 x 13
£22.40 +VAT

size 155 x 13
£20.60 +VAT

U-FRAME TYRE CENTRE
VAT Reg No 317 4362 01

QUANTITY	DESCRIPTION	UNIT PRICE	COST
4	Tyre155×13		
		VAT	
		TOTAL	

A1

4 Here are two advertisements for new exhaust systems:

Suppose you are looking for a new exhaust for a 1.6L Cortina. Which dealer would you use? Why? (VAT is 15%.)

━━━━━━━━━ WITH A FRIEND ━━━━━━━━━

5 A clothes shop has a mark-up percentage of 60%. VAT is 15%.
The cost price of a pair of jeans is £12.80. The shop assistant says that 75% (that is, 60% + 15%) must be added to the cost price to get the selling price:

[C] [1] [2] [.] [8] [0] [×] [.] [7] [5] [=] 9.60

So he works out that the selling price is £12.80 + £9.60 = £22.40. Decide between you why he is wrong. What is the actual selling price?

Calculating selling prices

1 Aysha works in a DIY store. The store prices articles like this:

STEP-LADDERS
£29.76 +VAT

When Aysha sells an article she has to calculate the selling price, including VAT at 15%.

a) To find the selling price of the step-ladders she presses:

[C] [2] [9] [.] [7] [6] [×] [1] [.] [1] [5] [=]

Write one or two sentences to explain why this gives the correct selling price.

b) Use Aysha's method to find the selling price of each of these to the nearest 1p:

2   In the sales, Harlequin Curtains reduces its prices by 18%.

a) Check that the sale price of the curtains is £29.93.

b) This is how the shop assistant calculates the sale price:

$$\boxed{C}\ \boxed{3}\ \boxed{6}\ \boxed{.}\ \boxed{5}\ \boxed{0}\ \boxed{\times}\ \boxed{.}\ \boxed{8}\ \boxed{2}\ \boxed{=}$$

Explain why this gives the correct result.

c) Use the shop assistant's method to find the sale price of each of these:

Curtain materi~~**l**~~ **... ...vet Curtains**

Was £~~7.84~~ per sq. yard   **NOW**

Were £~~138.50~~   **NOW**

3   An electrical shop buys washing machines for £250, and then adds a mark up of 60%. VAT is 15%. Which of these calculations will give the correct selling price?

A $\boxed{C}\ \boxed{2}\ \boxed{5}\ \boxed{0}\ \boxed{\times}\ \boxed{1}\ \boxed{.}\ \boxed{7}\ \boxed{5}\ \boxed{=}$

B $\boxed{C}\ \boxed{2}\ \boxed{5}\ \boxed{0}\ \boxed{\times}\ \boxed{.}\ \boxed{1}\ \boxed{5}\ \boxed{=}\ \boxed{\times}\ \boxed{.}\ \boxed{6}\ \boxed{0}\ \boxed{=}$

C $\boxed{C}\ \boxed{2}\ \boxed{5}\ \boxed{0}\ \boxed{\times}\ \boxed{.}\ \boxed{6}\ \boxed{0}\ \boxed{=}\ \boxed{\times}\ \boxed{.}\ \boxed{1}\ \boxed{5}\ \boxed{=}$

D $\boxed{C}\ \boxed{2}\ \boxed{5}\ \boxed{0}\ \boxed{\times}\ \boxed{1}\ \boxed{.}\ \boxed{6}\ \boxed{0}\ \boxed{=}\ \boxed{\times}\ \boxed{1}\ \boxed{.}\ \boxed{1}\ \boxed{5}\ \boxed{=}$

========= TAKE NOTE =========

Price without VAT is £4.76.
VAT is added at 15%.
Price with VAT = £4.76 × 1.15 ≈ £5.47

> 100% of £4.76
> +15% of £4.76
> = 115% of £4.76

Original price is £350.
There is 18% off all items in the sale.
Sale price = £350 × 0.82 = £287

> 100% of £350
> −18% of £350
> = 82% of £350

4   When you pay for an expensive item by cash, instead of by cheque or credit card, or when you buy a large number of items, a shopkeeper may offer you a *discount*. This is normally offered as a percentage, for example, a 5% discount.

How much would you actually pay if you bought:

a) £575 worth of lead crystal
b) £175 worth of lead crystal?

LEAD CRYSTAL END OF LINE

12% discount for orders **over £250**

A1

5 Employees at DIY Hardware Stores are given a 5% discount on everything they buy from the store.

How much would an employee pay for these two items? (One includes VAT, and the other does not. VAT is 15%, and is added before the discount is calculated.)

DIY £52.60 (including VAT)

DIY £17.80+VAT

6 Ann Jones is an agent for a home-shopping catalogue. She gets $12\frac{1}{2}$% discount on all the goods she buys from the catalogue.

Find the 'Total', 'Discount' and 'Amount due' for this order.

GOODS	COST
1 Electric kettle	£16.95
1 Lady's watch	£27.98
1 Acrylic blanket	£12.99
TOTAL	_____
DISCOUNT ($12\frac{1}{2}$%)	_____
AMOUNT DUE	_____

7 a) A Rolls Royce is for sale at £89000.
By paying cash you get a $7\frac{1}{2}$% discount. How much do you pay?

b) John Edmund bought a Mercedes for cash and paid £28000. The showroom price was £30000. What percentage discount was he given (to the nearest 1%)?

CORE

1   Here is a 3D drawing of a room containing
    a bed, a wardrobe, a table and a chair.

    Below is a *plan* view of the room.

    a) Which piece of furniture is missing?

    b) Copy and complete the plan.
       (Use dotted squared paper.)

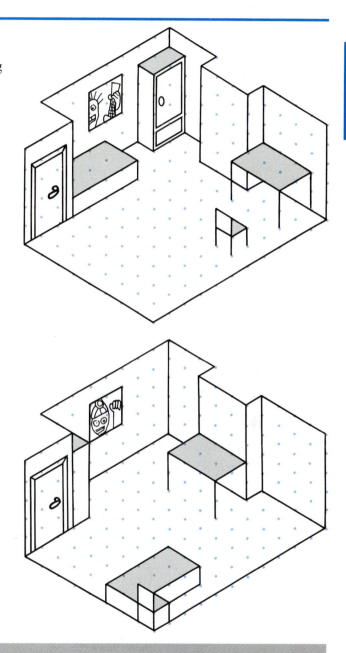

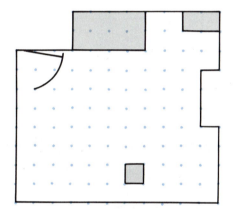

2   The furniture in question 1 has been
    rearranged like this. Draw the new plan.

CHALLENGE

3   Suppose you decide to use the room for a party. You want to have a lot of open floor space and
    you cannot remove any furniture. Draw a plan view of how you would rearrange the room.

A2

4 After your party you
   arrange the room like this.
   Make a 3D drawing of
   what the room looks like
   now. (Use dotted isometric
   paper.)

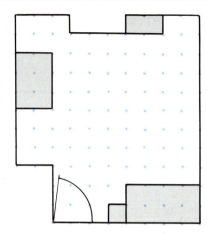

5 Six friends arrive to stay for the night.
   You remove the chair and table from the room and you
   add six more beds. Here is one way of fitting them in.

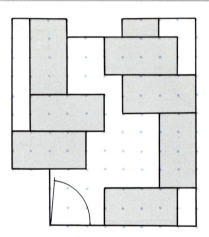

   Rearrange the furniture so that there is at
   least this much space between the sides of the beds.

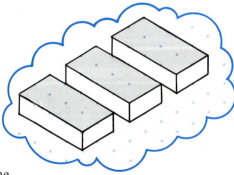

   Make sure the doors of the
   wardrobe and the room can still open!
   Draw a plan of your arrangement.

A2

6   This room contains a bed, a table, and a stool.

Copy and complete the plan of the room below.

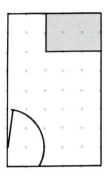

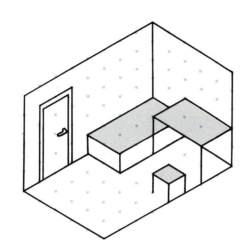

7   The room is rearranged as shown in the plan below:

Copy and complete the 3D view.

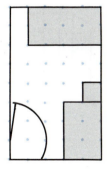

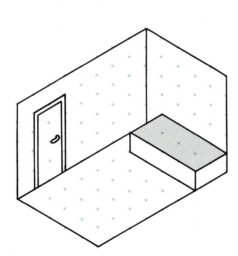

ACTIVITY

8   a) Draw a plan of how you would arrange the room in question 6.

b) Draw a 3D view of your plan.

A2

9   These rooms each contain a bed and a chest of drawers.
    Draw a plan of each room.

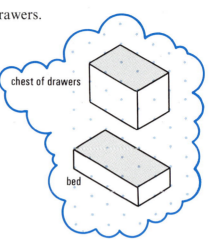

chest of drawers

bed

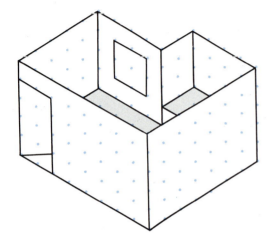

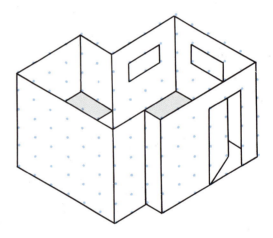

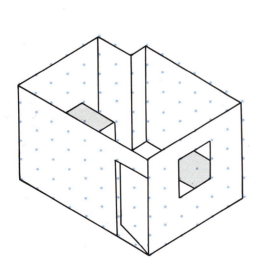

Garden sheds

1   A to H are garden sheds.

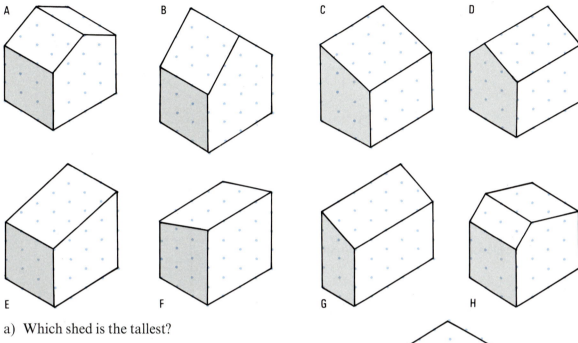

a) Which shed is the tallest?

b) One of the sheds does *not* fit on a base like this. Which is it?

c) This is a plan view of shed A. Which other shed has the same plan?

d) This is a plan view of shed C. Which other sheds have the same plan?

e) Draw the plan view of shed D. Use dotted squared paper.

THINK IT THROUGH

2   Draw another shed with the
same plan view as shed H in question 1.
Use dotted isometric paper.

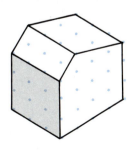

A2

3   We can look at shed A (in question 1) from
    direction X or from direction Y.

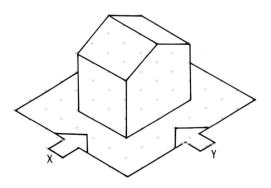

This is the *elevation* of shed A from direction Y.

Draw the elevation of shed B from direction Y.

4   This is the elevation of shed A from direction X.

   a) Two other sheds have this elevation from direction X.
       Which are they?

   b) Draw the elevation of shed B from direction X.

5   Look at these elevations
    from direction Y. Which
    sheds are they of?

6   a) Draw the elevation of shed C from direction Y.

    b) Which other sheds also have this elevation from direction Y?

━━━━━━━━━━━ THINK IT THROUGH ━━━━━━━━━━━

7   A shed has the elevation and plan shown,
    from direction Y.

    Draw its elevation from direction X.

ENRICHMENT

1   You will need dotted isometric paper and dotted squared paper. Choose a room at home and make a 3D drawing of it, like the one on page 7.

Show the position of the furniture, doors, windows and so on. To make the drawing you will have to imagine that two walls have been removed ...

... so make another drawing from the other 'end' of the room showing these two walls.

Finally, draw a plan of your room.

A2

A3 ESTIMATING

REVIEW

- 4291 is → 4290 rounded to the nearest 10
  → 4300 rounded to the nearest 100
  → 4000 rounded to the nearest 1000

■ Round 4291 to the nearest 5.

4735

4730   4740

- 4735 is 4740 rounded to the nearest 10 (for 'halfway' numbers we always 'round up').

■ Round 3750   a) to the nearest 100,   b) to the nearest 1000.

- By rounding up or down we can estimate results:

  £13.42   +£2.51   +£5.93

  ↗   ↑   ↖

  about £13   about £3   about £6   Estimate: about £22

■ In your head, estimate: a) £7.63+£4.80+£12.05   b) £17.64−£9.37.

CONSOLIDATION

1   a) In your head, estimate the total cost of each set of items:

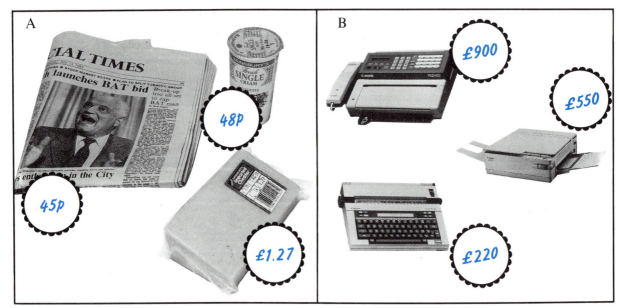

   b) Calculate the exact cost of each set of items in a).
      Check that your estimate gives a reasonable idea of the total cost. If it does not, check your
      calculation and your estimate.

2   a) In your head, estimate:

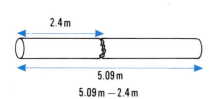

2.4 m

5.09 m

5.09 m − 2.4 m

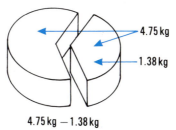

4.75 kg

1.38 kg

4.75 kg − 1.38 kg

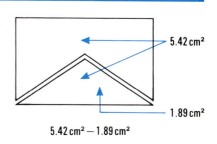

5.42 cm²

1.89 cm²

5.42 cm² − 1.89 cm²

b) Calculate the exact difference for each example in a).
Check that your estimate gives a reasonable idea of the difference. If it does not, check your calculation and your estimate.

3   Round 10 419   a) to the nearest 10   b) to the nearest 100   c) to the nearest 1000.

CHALLENGE

4   Sometimes ● rounding to the nearest 10
      and ● rounding to the nearest 5

give the same result. For example,   Sometimes they give different results. For example,

1279 is ⟶ 1280 to the nearest 10
        ⟶ 1280 to the nearest 5

1276 is ⟶ 1280 to the nearest 10
        ⟶ 1275 to the nearest 5

When do rounding to the nearest 5 and rounding to the nearest 10 give the same result, and when do they give different results?

CORE

Using estimates to check calculations

1   These are the numbers of female entrants in an annual marathon race recorded over a 10 year period:

1207   1715   2419   2097   3716   2984   3179   3467   3339   3098

a) Calculate the total number of entrants recorded over the 10 years.

b) How do you know that your result is correct?
If you do know, explain how you know. If you are not certain, explain how you can check.

c) By adding the number of entrants again, this time starting with the last number, 3098, check your result in a). If your results do not agree, check by adding the numbers in a different order.

It is always useful to have a rough idea of a calculator result *before* you do the calculation.

Four thousand nine hundred and sixteen and nine thousand three hundred and twenty... that's about fourteen thousand.

My calculator gives 140236... that can't be right... I must have pressed a wrong key...

4916 + 9320

About 5000   About 9000

Estimate : About 14000

If your estimate and calculation do not roughly agree, you can then make a check (of both your estimate *and* your calculation).

A3

2   a) Round each number   (i) to the nearest 100,   (ii) to the nearest 10:

   4196   375   107   91   38

   b) Use your approximations in a) to estimate, in your head, the result of each of the calculations below. Use whichever approximation helps you to get a quick reasonable estimate:

   (i) 4196 + 375   (ii) 375 + 38 + 107   (iii) 375 − 91   (iv) 4196 − 375.

   c) Find the exact results in b).
      Check that your estimates give a reasonable idea of the results. If they do not, check your calculations and your estimates.

3   The table shows the sales figures in 1975 and 1985 of a local newspaper.

Sales figures	
1975	*1985*
15 476 504	18 307 406

Calculate how many more papers were sold in 1985 than in 1975. Don't forget to make an estimate first.

4   a) Each set of three calculator displays shows three attempts at a calculation. In each case, decide which calculation is most likely to be correct, by making a rough estimate in your head:

   (i) 17 406 + 3949 | 2135. | | 21355. | | 213550. |

   (ii) 4916 − 2098 | 2818. | | 28108. | | 7014. |

(iii) $4073 + 359 - 3190$

(iv) $(219 + 98) - (37 + 89)$

b) Find the exact result for each calculation in a).
   Check that the estimate you made for each one in a) was reasonable. If it was not, check your estimate and your calculation.

Estimates for decimals

1 a) Copy the number line:
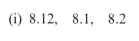

b) Using arrows (as for 0.8 and 1.4) mark approximately the positions of:
   0.38   1.27   2.7

▨▨▨▨▨▨▨▨▨▨▨▨ TAKE NOTE ▨▨▨▨▨▨▨▨▨▨▨▨

From the number line we can see that 1.4 is nearer to 1 than it is to 2. We say that:

- 1.4 is *1 to the nearest whole number*
- 7.6 is *8 to the nearest whole number*
- 3.5 is *4 to the nearest whole number* (we always 'round up' for 'halfway' numbers)

c) Write 0.38, 1.27, 2.7 and 3.15 to the nearest whole number.

2 a) Write each trio of numbers in order, smallest first. The number lines will help you.

   (i) 8.12,   8.1,   8.2
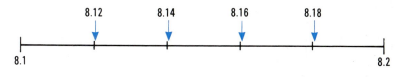

   (ii) 8.15,   8.143,   8.14
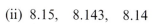

b) Copy the number line.
   Mark on it a number
   which lies between
   8.144 and 8.145.
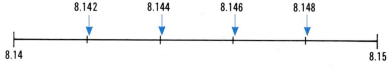

c) Write down a number which lies between 8.144 7 and 8.144 8.

3　Explain why 8.114 916 4 is not larger than 8.12.

===== THINK IT THROUGH =====

4　Write each of these to the nearest whole number:

a) 10.779 8　　　b) 4.765　　　c) 3.333 333 3　　　d) 9.019 1

e) | 171.746789 |

f) | 0.008399978 |

g) | 4.44444444 |

A3

===== CHALLENGE =====

5　a)　Write each trio of numbers in order, smallest first:

(i)　7.12,　7.125,　7.122
(ii)　7.12,　7.122,　7.121 5
(iii)　7.121 57,　7.121 58,　7.121 577
(iv)　7.121 577 894,　7.121 577 88,　7.121 577 819 84.

b)　Write down a number which lies between:

10.019 841 766 517 641 476 and 10.019 841 766 518 641 476.

6　a)　Write each of these to the nearest whole number:

(i) 2.4　(ii) 3.97　(iii) 61.80　(iv) 27.02　(v) 137.614.

b)　Write each number in a) to the nearest 10.

c)　Use your results from a) and b) to estimate, in your head:

(i)　　　2.4 + 61.80
　　　　　↗　　　↖
　　　about 2　　about 60

(ii) 61.80 − 2.4　(iii) 137.614 + 61.80　(iv) 27.02 − 3.97

(v) 3.97 + 2.4 + 27.02　(vi) 137.614 − 61.80.

d)　Use your calculator to find the exact results in c). If your calculator result and your estimate are not approximately the same, check each one.

▓▓▓▓▓▓▓▓▓▓▓▓▓▓ TAKE NOTE ▓▓▓▓▓▓▓▓▓▓▓▓▓▓▓▓▓▓▓▓▓▓▓

To estimate results ● we sometimes round up
● we sometimes round down
● we sometimes round to the nearest whole number
● we sometimes round to the nearest 10, nearest 100, …

We choose the approximation for each number which will give us a reasonable result. We also choose the approximations to make it easy to do mental estimates.

Here are some examples:

$0.424 + 1.176$

↗   ↖
about 0.4 + about 1.2   Estimate: about 1.6

$76.76 - 43.62$

↗   ↖
about 80 − about 40   Estimate: about 40

$17.8 + 1764.6$

↗   ↖
about 20 + about 1760   Estimate: about 1780

$1764.6 + 9439.8$

↗   ↖
about 2000 + about 10 000 Estimate: about 12 000

A3

7   Here are the results of some calculator additions and subtractions. Decide which calculation is most likely to be correct for each one:

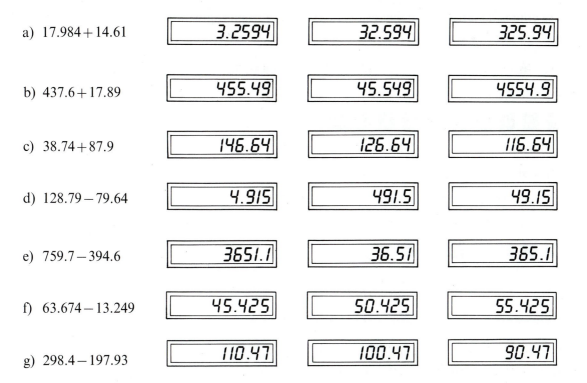

a) $17.984 + 14.61$ 3.2594   32.594   325.94

b) $437.6 + 17.89$ 455.49   45.549   4554.9

c) $38.74 + 87.9$ 146.64   126.64   116.64

d) $128.79 - 79.64$ 4.915   491.5   49.15

e) $759.7 - 394.6$ 3651.1   36.51   365.1

f) $63.674 - 13.249$ 45.425   50.425   55.425

g) $298.4 - 197.93$ 110.47   100.47   90.47

ENRICHMENT

1   a)   In your head, estimate each of these (round each number to the nearest whole number, then do the calculation):

   (i) $7.68 \div 4.02$ (ii) 0.74×19.6 (iii) 11.074×13.92
   (iv) $26.7 \div 14.85$ (v) $83.66 \div 9.98$ (vi) 3.6×4.12
   (vii) 8.042×4.67 (viii) $19.6 \div 4.04$ (ix) $24.3 \div 7.86$.

   b)   Find each result in a) using your calculator. Check that your estimates and calculator results roughly agree.

2   a)   Check that 27.76 is 30 to the nearest 10.

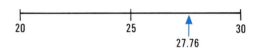

   b)   Round each of these to the nearest 10:

   (i) 19.4   (ii) 8.76   (iii) 75.9   (iv) 94.83   (v) 103.76.

   c)   Use your results in b) to estimate each of these:

   (i) 19.4×8.76 (ii) 75.9×103.76 (iii) $94.83 \div 8.76$.

   d)   Find each result in c) using your calculator. Check that your estimates and calculator results roughly agree.

3   a)   In your head calculate:

   (i) $7 \div 3 \times 3$ (ii) $15 \div 6 \times 6$ (iii) $1 \div 9 \times 9$ (iv) $23 \div 6 \times 6$.

   b)   Use your calculator to find the results in a). Do the results agree with your mental calculation? If not, can you explain why?

═══════ CHALLENGE ═══════

4   a)   Key into your calculator

   [0] [.] [3] [3] [3] [3] [3] [3] [3] [3]

   (as many 3s as your calculator display will take).

   Multiply by 3. Write down the result.

   b)   In your head, calculate $1 \div 3 \times 3$.

   c)   John entered on his calculator

   [C] [1] [÷] [3] [×] [3] [=]

   His calculator gave the result: `0.99999999`

   Is this correct? If you say 'No', explain why you think his calculator gave the wrong result.

REVIEW

The *square root* of 36 is 6, because $6 \times 6 = 36$. We write $\sqrt{36} = 6$.

Three squared is written as 3^2. $3^2 = 3 \times 3 = 9$

Two cubed is written as 2^3. $2^3 = 2 \times 2 \times 2 = 8$

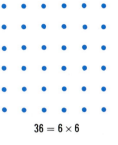

$36 = 6 \times 6$

[C] [4] [.] [2] [√] [=] displays the square root of 4.2.

[C] [4] [.] [2] [x^2] [=] displays the square of 4.2.

■ What is the value of $2^3 \times 4^2 \times \sqrt{16}$?

CONSOLIDATION

IN YOUR HEAD

1   Do these in your head. Write down only the answer.

a) Write down the value of: (i) 2×2 (ii) 3^2 (iii) $\sqrt{4}$ (iv) 4^2.

b) Write down the value of: (i) four squared   (ii) two cubed   (iii) the square root of 49.

c) Find the length
of side of each square:

Area 16 cm²

Area 81 cm²

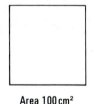

Area 100 cm²

d) Write down the value of:

(i) 0×0 (ii) 1×1 (iii) 0^2 (iv) 1^2 (v) $0 \times 0 \times 0$ (vi) 0^3 (vii) $1 \times 1 \times 1$ (viii) 1^3.

e) *Imagine* you press these sets of keys on your calculator. Write down the result you would get.

(i)   [C] [2] [x^2]   (ii)   [C] [8] [1] [√]   (iii)   [C] [2] [x^2] [√]

f) Use $\sqrt{}$ to write the   (i)   (ii)
length of the side of each square:

Area 26 cm²

Area 319 cm²

g) Write down the value of:

(i) $2^2 \times 2$ (ii) $\sqrt{25} \times 4$ (iii) $3^2 \times 2$.

A4

1   Do these in your head. Write down only the answer.

a) The photograph is a square of area 64 cm^2.

What is the length of each side?

b) Write down the value of:

(i) the square root of 81   (ii) three cubed   (iii) 5^2 (iv) $\sqrt{16}$
(v) one squared   (vi) the square root of zero.

c) Estimate $\sqrt{90}$ to the nearest whole number.

d) Write down the whole number whose cube lies between 60 and 70.

e) The volume of this wooden cube is 30 cm^3. How long is each
    side to the nearest centimetre?

f) Suppose you press these keys on your calculator:

What result do you get?

g) What result do you get when you press these keys on your calculator?

━━━━━━━ WITH A FRIEND ━━━━━━━

2   Here are two games – the *square root* game and the *squares* game. Play each one with your friend.

GAME 1 The square root game

Player 1
- chooses a whole number from 1 to 20, ○ ○ (*e.g. seventeen*)

Player 2
- guesses the square root, ○ ○ (*e.g four point seven*)
- checks the guess *using only the* $\boxed{\times}$ *key* (or $\boxed{x^2}$ *key*),

$\boxed{C}$ $\boxed{4}$ $\boxed{.}$ $\boxed{7}$ $\boxed{x^2}$ $\boxed{=}$ $\boxed{ 22.09}$

- has a final guess and records it: 4.2 ○ ○ (*four point two*)
- checks this guess using only the $\boxed{\times}$ or $\boxed{x^2}$ key.

$\boxed{C}$ $\boxed{4}$ $\boxed{.}$ $\boxed{2}$ $\boxed{x^2}$ $\boxed{=}$ $\boxed{ 17.64}$

Now *player 2* chooses a number, and *player 1* guesses and checks twice, as above.

Score: The one whose final guess is nearer to the true value of the square root scores 1 point (subtract the final guess from the true square root to decide who is nearer). Repeat the game until one player has 5 points. 5 points wins!

GAME 2 The squares game

Player 1
- chooses a number between 20 and 40.

Player 2
- has two guesses to find the square, and checks *using only the* $\boxed{\sqrt{}}$ *key*.

Now *player 2* chooses a number, and the rest of the game is played like Game 1.

Score: The first player to score 5 points wins.

A4

━━━━━━━ EXPLORATION ━━━━━━━

3   Look at the flow diagram. Investigate what happens for other 2-digit numbers. Do you always arrive at a whole number result? Write a report about what you discover.

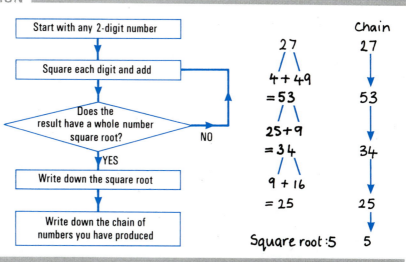

ENRICHMENT

1 Find x in each sentence, by estimating and checking:

a) $\sqrt{x} = 81$ b) $x^2 = 1$ c) $4^x = 64$ d) $x^2 = 121$

e) $x^3 = 0$ f) $\sqrt{x} = 1$ g) $x^2 + x = 6$ h) $x^2 = 0.64$.

2 a) Do *not* use a calculator in this part.
      Which is larger: (i) 0.1 or 0.1^2 (ii) $\sqrt{0.9}$ or $\sqrt{0.4}$?

    b) Check your results in a) with a calculator.

3 a) Do *not* use a calculator in this part.

      (i) Which of these is nearer to $\sqrt{0.9}$:   A 0.3   B 0.4   C 0.9?

      (ii) Which of these are whole numbers: A $\sqrt{54}$ B $\sqrt{1000}$ C $\sqrt{324}$ D $\sqrt{484}$?

    b) Check your results in a) with a calculator.

4 Using only the ☒ = ▪ and digit keys on a calculator, find the missing numbers:

a) $\square^3 = 54.872$ b) $\sqrt{566.44} = \square$

A4   ▬▬▬▬▬ CHALLENGE ▬▬▬▬▬

5 Use only the ☒ = ▪ and digit keys on a calculator.

By guessing and checking find $\sqrt{12.3}$ correct to the nearest 0.01.

6 Without using a calculator, find the value of:

a) $\sqrt{10} \times \sqrt{2} \times \sqrt{10} \times \sqrt{2}$

b) $\sqrt{7} \times 3^2 \times \sqrt{7}$

c) $\dfrac{\sqrt{4} \times \sqrt{4}}{\sqrt{2} \times \sqrt{2}}$

CORE

Wages and salaries

================ WITH A FRIEND: WAGE CHALLENGE ================

1 When you work you can be paid in various ways:

per hour	for example,	£3.75 per hour
per week	for example,	£420 per week
per month	for example,	£1200 per month
per annum (year) or p.a.	for example,	£15 000 per year.

Assuming that for each of the rates above, the number of hours work per day is 8, there are 4 weeks in a month, and there are 11 of these working months in a year, which rate of pay is best?

================ TAKE NOTE ================

Working hours are often fixed; for example, 9:00 a.m. to 5:15 p.m. with an unpaid hour for lunch. If you work more than these hours, this is called *overtime*. Normally you are paid more than the standard rate for overtime.

For example, Ahmid works in a timber yard.
His working day is 8:00 a.m. to 5:00 p.m. with a one hour unpaid lunch break (that is, an eight hour day). He often works two hours *overtime* from 5:00 p.m. to 7:00 p.m.

His *standard* pay rate is £3.60 per hour.
His *overtime* pay rate is £4.90 per hour.

A5

2 Read the *Take note*.

a) How much does Ahmid earn per day when he does two hours overtime?

b) This is the work record card for Ahmid in one week.

How much did Ahmid earn for the week?

	STANDARD				OVERTIME	
	a.m.		p.m.			
	Start time	Finish	Start time	Finish	Start time	Finish
Monday	8:00	12:00	1:00	5:00	5:00	6:30
Tuesday	8:00	12:00	1:00	5:00	—	—
Wednesday	8:00	12:00	1:00	5:00	—	—
Thursday	8:00	12:00	1:00	5:00	—	—
Friday	—	—	1:00	5:00	5:00	7:00

3   Wage earners are paid at the end of every week. Salary earners are normally paid every month. Bus drivers, bricklayers and shop assistants are usually wage earners. Teachers, nurses and accountants normally earn salaries.

Collect some examples of job advertisements from the newspapers. Compare the amounts people earn.

For example, find advertisements for people who will earn:

more than £50 000 per year
between £30 000 and £50 000
between £20 000 and £30 000
between £15 000 and £20 000
between £10 000 and £15 000
between £7500 and £10 000
between £5000 and £7500

Describe how the working hours, conditions and benefits (such as company car, medical insurance, staff discount, mortgage assistance) differ from one type of work to another.

Choose examples of different types of jobs and different types of payments (salaried, wage-earning, part-time, etc.).
Compare the examples you have collected by working out how much you think each person earns per hour (taking into account all the 'fringe benefits').
Write a report about what you discover

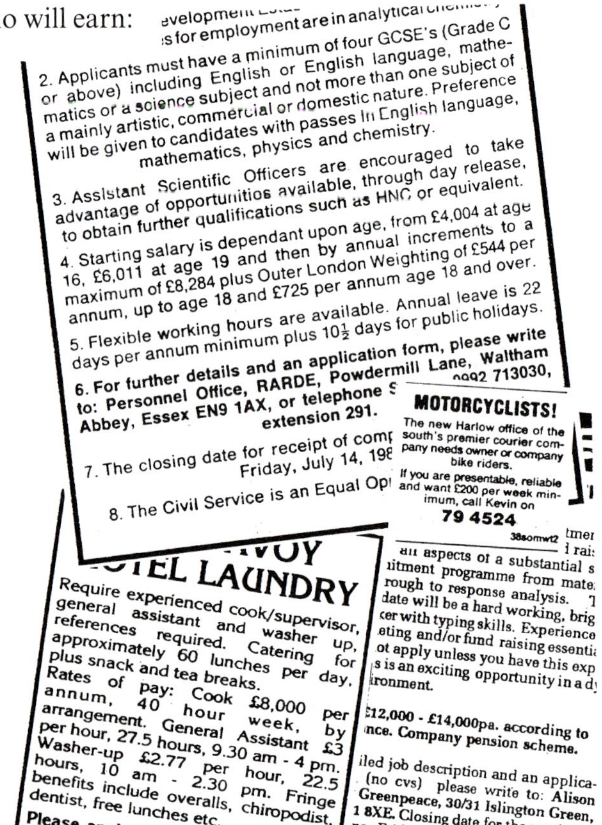

A5

Overtime is often paid at '*time and a half*'. For example, if the standard rate is £3.60 per hour the overtime rate is £5.40 per hour (£3.60 + $\frac{1}{2}$ of £3.60).

Weekend (or Sunday) work might be paid at '*double time*'.
When the standard rate is £3.60 per hour the double-time rate is £7.20 per hour (2 × £3.60).

---

4   Read the *Take note.*

a) Mary works
6 hours for £3.60 an hour,
2 hours at time and a half,
and 1 hour at double time.

How much does she earn?

b) Rula has a part-time job.
The standard rate is
£2.80 per hour.

This is her time card for
the week.

Calculate how much she
earns.

	STANDARD (hours)	TIME AND A HALF (hours)	DOUBLE (hours)
Mon	5	–	–
Tues	4	2	–
Wed	–	–	–
Thurs	2	–	–
Fri	4	1	–
Sat	3	–	–
Sun	–	–	2

A5

Tax and other deductions

1   All of us have to pay income tax on earnings over a certain limit. We all pay National Insurance and most people contribute to a company pension scheme.

a) Gerald earns £247 per week before any deductions. This is called his *gross* pay. The first £53 is tax free. He pays 25% tax on the remainder. How much does he pay in tax each week?

b) For National Insurance Gerald pays £22.23 per week. There are no other deductions from his wages. How much does he take home each week in his pay packet? (This is called his *net* pay).

2   This is Helen's payslip for the 12th working week of the year.

Tax and pay year to date	Employee's contributions	Name: H Street Staff No. 7–0312 Period 12	
Code No. 203 L	**National Insurance** 73.44	**Gross pay and additions**	**Deductions**
Gross pay 816.00	**Pension** —	Gross pay 68.00	Tax 8.63 Insurance 6.12
Taxable pay 470.64			
Tax 103.56		Gross pay 68.00 Net pay 53.25	Total deductions 14.75 Company name BLAKE ASSOCIATES

a) From the payslip you can see that she has earned gross pay of £816.00 (in the 12 weeks). Check that this gives £68.00 per week.

b) How much does she pay each week in National Insurance contributions?

c) What percentage of her gross pay is take-home (net) pay?

3

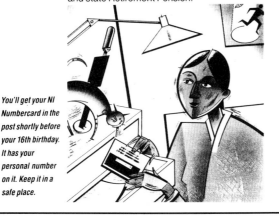

What is National Insurance?

Most people who are between 16 and state pension age (60 for women and 65 for men) and who work must pay contributions into the National Insurance (NI) scheme. These contributions enable you to get NI benefits such as Unemployment Benefit, Sickness Benefit, and state Retirement Pension.

You'll get your NI Numbercard in the post shortly before your 16th birthday. It has your personal number on it. Keep it in a safe place.

Your NI Numbercard

You should have received your NI Numbercard in the post shortly before your 16th birthday. It has your personal NI number printed on it. Keep it in a safe place. Your employer will need to know your number so that your NI contributions can be put on to your own contributions record.

Your Social Security office will also need to know your number if you need to claim benefits.

If you can't find your Numbercard or have never had one, ask your Social Security office for one straight away.

Your NI contributions

If you work for an employer

If you work for an employer and earn at least a set amount, you must pay **Class 1** NI contributions. The more you earn the more you pay. Your employer takes your contributions out of your pay and also has to pay employers' contributions for all employees who earn at least the set amount.

a) In 1989, your National Insurance contributions would have been 9% of your gross pay. How much would you have paid on a gross salary of £12 700 per year?

b) As a single person the income tax you pay normally increases in bands like this:

Band 1	Up to £3500	No tax
Band 2	Basic rate: Next £14 500	28%
Band 3	Higher rate: Next £10 000	35%
Band 4	Super rate: Over £28 000	40%

The tax bands and rates change year by year in the Budget so that the Government can collect more, or less, revenue for spending on roads, education, health, defence, etc. How much tax would you pay on a salary of £20 000, using these tax bands?

ENRICHMENT

1   These were the tax bands and the tax rates in 1986–87:

29% basic rate	First £17 200 of taxable income
40% rate	Next £3000
45% rate	Next £5200
50% rate	Next £7900
55% rate	Next £7900
60% rate	Anything more

How much income tax would a company director with a taxable income of £50 000 have paid in 1986–87?

ASSIGNMENT

2   Find out the present tax bands and rates. Calculate how much tax a company director with £50 000 of taxable income would pay.
Would she/he pay more today than in 1986/87? Explain why you think this is so.

A5

REVIEW

Number of eggs in 30 rooks' nests

4	3	3	1	2
2	3	4	2	4
4	2	1	4	3
3	3	3	2	4
0	4	1	0	2
3	4	2	2	4

The data about the number of eggs in 30 nests can be represented in:

- A frequency table:

Number of eggs	Tally	Frequency
0				2
1					3
2	⊬⊬ |||	8
3	⊬⊬ |||	8
4	⊬⊬ ||||	9
	TOTAL	30

- A variety of types of diagram, for example:

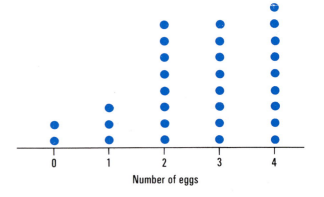

■ What other kinds of diagram could be used to represent the data? Name two.

- The most common number of eggs is 4. This is the *modal value* or *mode* of the distribution.

- When the number of eggs are arranged in order:

  0, 0, 1, 1, 1, 2, 2, 2, 2, 2, 2, 2, 3, 3, 3, 3, 3, 3, 3, 3, 4, 4, 4, 4, 4, 4, 4, 4, 4

  the 'middle value' is called the *median*.
  Here, it is the mean of the 15th and 16th values, that is, $\frac{3+3}{2} = 3$.

- The *mean* of the distribution tells us how many eggs would be in each nest if the eggs were 'shared out equally'.
  There are 79 eggs.
  The mean is therefore $79 \div 30 \approx 2.6$.

■ What are the mode, median and mean of these five distances: 3 cm, 3 cm, 4 cm, 5 cm, 15 cm?

CONSOLIDATION

━━━━━━━━━━━━━ WITH A FRIEND ━━━━━━━━━━━━━

1 a) Discuss together what
      the basketball coach
      means. Write down what
      you decide.

   b) Make up a cartoon of
      your own which
      uses mean, mode
      and median.

'Should we scare the opposition by announcing our mean height or lull them by announcing our median height?'

2 Write down a set of five ages for which the mean is 5 years, the mode is 2 years and the median is 3 years. There are many different answers. Find just one solution.

3 a) Find the mean weight of each set of potatoes:

      (i) 200 g, 22 g, 400 g   (ii) 500 g, 200 g, 100 g   (iii) 50 g, 50 g, 700 g.

   b) Write one or two sentences to explain why the mean does not always give us a good idea of actual measurements. Part a) will help you.

4 The bar charts show the number of beagle pups in each litter, for two different kennels during one year.

   a) How many litters were produced in each kennel during the year?

   b) How many pups were born in each kennel during the year?

   c) What do you think is the most common size of beagle litter?

   d) What is the mean size of a beagle litter in each kennel?

   e) What is the median size of a beagle litter in each kennel?

   f) When the next litter of pups is born, what do you think is the most likely number of pups in each kennel? Give your reasons.

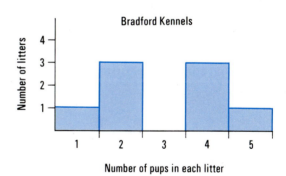

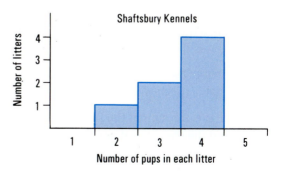

A6

Surveys

1 Thirty people were asked 'How many times did you visit your doctor last year?' The grid shows all the responses.

0	5	1	3	4
8	12	0	1	2
3	2	0	0	6
10	0	9	5	15
1	0	3	2	0
16	1	4	2	3

a) Copy and complete the frequency table:

Number of visits	Tally	Frequency
0 - 4	ⳠⳠⳠⳠⳠ	
5 - 9	I	
10 - 14	I	
15 - 19		

b) Copy and complete the distribution diagram to show the information in the grid:

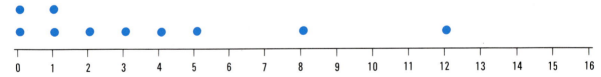

A6

Check that the frequencies in your frequency table and on your distribution diagram agree with each other.

c) (i) Which number of visits to the doctor was most common?
   (ii) What is the *mode* of the distribution?

d) (i) Arrange the number of visits in order, like this: 0, 0, 0, 0, 0, 0, 0, 1, 1, 1, 1, 2, 2, ...
   (ii) What is the *median* value of the distribution?

e) (i) What is the sum of the frequencies in your table?
   (ii) Find the *mean* number of visits per person.

━━━━━ TAKE NOTE ━━━━━

The number of visits made *range* from 0 to 16.
We say that the *range* of the distribution is $16-0 = 16$.
The range gives us an idea of how widely the data is spread out.

━━━━━ THINK IT THROUGH ━━━━━

f) The grid shows the data from a survey of another 30 people for the same doctor.

2	1	0	0	1	2
3	1	0	0	3	4
1	1	0	4	9	4
20	1	2	2	2	1
3	2	1	0	1	2

 (i) Draw a distribution diagram and calculate the mean, mode, median and range.
 (ii) Compare your results from the two surveys. What do they suggest about the two groups of people? Write one or two paragraphs to explain.

2 This is a *bar-line* graph. It shows the sales of various sizes of shoes in a children's shoe shop during one Saturday.

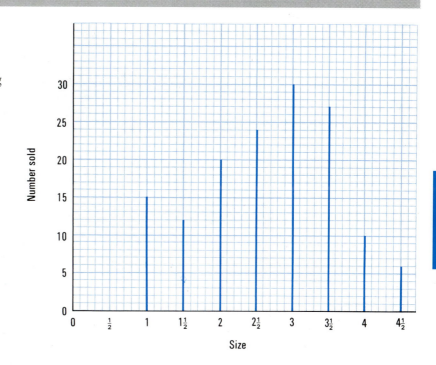

a) How old do you think most children were for whom shoes were bought?

b) (i) What was the most common size bought?
   (ii) Is the value in (i) the mean value, the modal value or the median value?

c) What is the range of the distribution?

d) Imagine you draw a bar-line graph for the sizes of girls' shoes bought, and one for the sizes of boys' shoes bought. How would you expect:

 (i) the range (ii) the mode (iii) the median (iv) the mean to differ from graph to graph? Why?

A6

ASSIGNMENT: SURVEYS

3 a) Think of a survey, like those in questions 1 and 2, which you can carry out. Write down the question for your survey and collect responses from at least 20 people. Draw a distribution diagram (dot graph or bar-line graph), and find the mean, mode, median and range for your distribution.

   b) Think of carrying out your survey on a different set of people to those you asked. Explain how you would expect the results from your two surveys to vary, and why.

4 a) Think of two different groups of people (for example, children and parents; first years and fourth years). Think of a question you would like to ask the two groups, so that you can make a comparison. Before you carry out your survey write down how you think the results will vary. Say how you think the mean, mode, median and range will vary.

   b) Carry out your survey on each group of people. Draw a frequency diagram and a distribution diagram. Write a report to compare the results from the two groups. Were the results similar to what you imagined in a)?

ASSIGNMENT

5   You should collect data for this assignment over a period of about four weeks. As you receive coins for change, for pocket money, etc., keep a note of the date of each one. Keep records of the dates for each type of coin, 1p, 2p, 5p, 10p, 20p, 50p …

Draw charts to show the dates for each type of coin, and work out the mean, median, and modal dates and the range of dates. Choose the types of charts which you think illustrate your information in the best possible way.

Write a report about what you discover. In your report say what your results suggest about the numbers and dates of each type of coin in circulation. (For example, your results might suggest that more 1ps were made in 1982 than in any other year, and so on.)

A7 SPECIAL NUMBERS

REVIEW

- 1, 2, 3, 4, 6 and 12 are *factors* of 12.

- 12 is a *multiple* of 1, 2, 3, 4, 6 and 12.

  The first five multiples of 4 are 4, 8, 12, 16, 20.

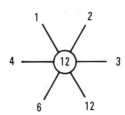

■ What are the factors of 18? What are the first three multiples of 18?

- The first five *square* numbers are:

$1 = 1 \times 1$ $4 = 2 \times 2$ $9 = 3 \times 3$ $16 = 4 \times 4$ $25 = 5 \times 5$

■ What is the tenth square number?

- The first five *prime* numbers are: 2, 3, 5, 7, 11.
  Prime numbers have exactly *two* factors, 1 and the numbers themselves:

$1 — ② — 2$ $1 — ③ — 3$ $1 — ⑤ — 5$ $1 — ⑦ — 7$ $1 — ⑪ — 11$

Note: 1 is *not* a prime number since it has not got two distinct factors.

■ What is the tenth prime number?

- The first five triangle numbers are: 1   3   6   10   15

■ What is the tenth triangle number?

CONSOLIDATION

IN YOUR HEAD

1   Do these in your head. Write down only the answers.

- a) The first multiple of 7 is 7. What is the seventh multiple of 7?
- b) 18 has four factors not counting 1 and 18. What are they?
- c) 22 and 55 share two factors. What are they?
- d) Which number has only one factor?
- e) What is the first number which is both a square number and a triangle number?
- f) Write down the first number which is a multiple of both 4 and 6.
- g) The tenth triangle number is 55. What is the eleventh?
- h) Add together the first four odd numbers. Which square number is this?

2   Copy and complete each factor star.

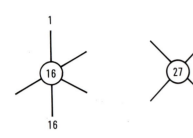

3   Write down the next number in each sequence:

a) 2, 4, 6, 8, 10, ...

b) 2, 4, 8, 16, 32, ...

c) 1, 3, 5, 7, 9, 11, ...

d) 2, 3, 5, 7, 11, 13, ...

e) 1, 4, 9, 16, 25, 36, ...

 EXPLORATION

4   a)   These are multiples of 7:   7, 14, 21, 28, 35, 42, ...
                                       ↓   ↓   ↓   ↓   ↓   ↓
         These are their digit sums: 7,   5,   3, 10,   8,   6, ...
                                                        ↓
                                       $1+4$ 1

         This is a diagram which represents the digit sums:

         Copy and complete the diagram.

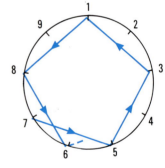

   b) Draw a diagram like that in a) for the digit sums of
       multiples of 5.

   c) A and B are diagrams for the digit sums of the multiples of
       two other numbers. Which numbers are they?

   d) Investigate some examples of
       your own. What different
       kinds of diagrams can
       be produced?

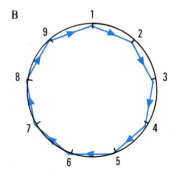

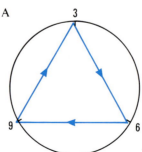

CORE

Division checks

1 a) Write down four numbers which are divisible by 2.
What special kinds of numbers are divisible by 2?

   b) Write down the first six numbers which are divisible by 5.
We can tell immediately when a number is divisible by 5 by looking at the units digit:

   1245   13 990   14 552   17 963
   YES!   YES!   NO!   NO!

   What do you notice about the units digits of numbers divisible by 5?

   c) What is special about numbers which are divisible by 10?

CHALLENGE

2 a) Check that these numbers are divisible by 4:

   (i) 41 212 (ii) 341 288 (iii) 73 080   (iv) 916 964.

   b) We can tell quite quickly which numbers are divisible by 4 by inspecting the last two digits:

   17 792   41 374   41 372   71 450
   YES!   NO!   YES!   NO!

   Investigate what is special about the last two digits of numbers which are divisible by 4. Write down what you discover.

EXPLORATION

3 a) We can tell quite quickly which numbers are divisible by 3 by adding the digits:

   121 741   314 961   400 443   17 192
   16 NO!   24 YES!   15 YES!   20 NO!

   Investigate some examples of your own. Explain how we can tell if a number is divisible by 3, by adding the digits.

   b) Investigate which numbers are divisible by 9.
How can you tell quite quickly if a number is divisible by 9 or not?

   c) Investigate some other numbers used as divisors, for example, 11, 15, 20, ... Find some quick 'division checks' of your own.

A7

ENRICHMENT

THE COMMON FACTOR GAME

1 Play this game with a friend.

- Each player, in turn, thinks of a number between 500 and 1000 and writes it down.
- The one who chose the first number now has one minute to find a number other than 1 which is a factor of *both* of the numbers chosen.
- After one minute the second player tries to find another 'common factor'.
- When neither player can find another common factor, the game ends. Score one point for each factor found.
- Choose two more numbers and repeat the game.

The first player to score ten points wins the game.
Note: '1' is *not* allowed as a factor in this game.

A7

REVIEW

● A circle folds exactly in half along a diameter.

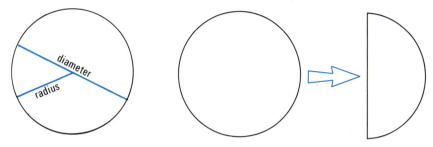

Diameter = 2 × Radius

■ How can you find the centre of a paper disc, by folding?

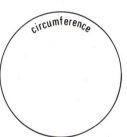

● The distance around a circle is called its *circumference*.

Circumference ≈ 3 × Diameter

■ The wheels of a bicycle have a diameter of 60 cm. Roughly, how far does the bicycle travel for each turn of its wheels?

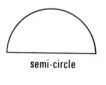

semi-circle

quadrant of a circle
(quarter of a circle)

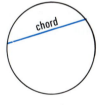

chord

arc

40° sector

120° sector

■ What is the angle of each sector you get when you fold a circle in half, and then keep halving, like this?

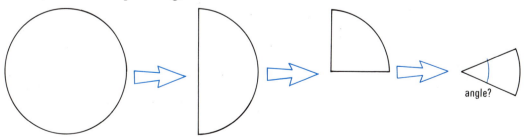

angle?

CONSOLIDATION

1   Use Circumference $\approx 3 \times$ Diameter to do these:

a) Bicycle wheel, diameter 66 cm.
   What is the length of a spoke
   (approximately)?

b) Paint can, radius 11 cm.
   What is its circumference?

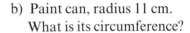

c) Earth, circumference 40106 km.
   What is the radius?

2   The label is for the soup can. Roughly, how long
   and how wide should it be?

3   Use Circumference $\approx 3 \times$ Diameter in each part.

a) Find the *circumference* of each of these circles:

(i)

(ii)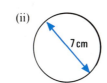

(iii)

b) Find the *diameter* of each of these circles:

(i)

(ii)

(iii)

c) Find the *radius* of each of these circles:

(i)

(ii)

(iii)

CORE

CHALLENGE

1   a) These circle patterns are made from wire. Which one uses most wire?

A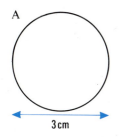

3 cm

B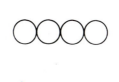

3 cm

C

3 cm

D

3 cm

b) These curves are all made from semicircles.

A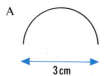

3 cm

B

3 cm

C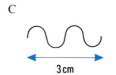

3 cm

D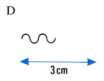

3 cm

Make your own complete sketch of D. Which curve is longest?

A8

TAKE NOTE

An approximation for the *circumference* of a circle is   3 × Diameter.
A better approximation is   3.1 × Diameter
An even better approximation is   3.14 × Diameter

2 a) Use each of the three approximations in the *Take note*
       to find the circumference of this
       storage tank:

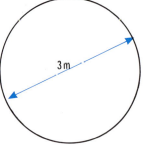

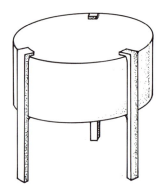

3 m

   b) What is the difference in
       centimetres between the
       greatest distance and
       the least distance given
       by the three
       approximations?

   c) The side of the storage tank is made of sheet metal.
       What length of sheet metal would you order?

3 a) Use the three approximations in the
       Take note to find the distance around
       the equator of the Earth, given that the
       diameter of the Earth is 12 766 km.

   b) What is the difference in distance given
       by the least accurate and most accurate
       approximations?

A8

TAKE NOTE

The *exact* circumference of a circle is Diameter × 3.141 592 653 589 793 238 462 643 383 279 502 88 …

(This string of digits never ends, and it does not have a repeating pattern.)

It is impossible to write down the exact number which multiplies the diameter. We would be writing forever! So instead we call the exact number π (the Greek letter 'pi'), and so

Circumference = Diameter × π
  or = π × Diameter

For most calculations we use 3.1 or 3.14 or $\frac{22}{7}$ for π.

4 Check to see if your calculator has a ▮π▮ key. If it does, press it. What value does your calculator display for π?

5 a) Use your calculator to find the value of $\frac{22}{7}$ as a decimal. How many digits after the decimal point are identical to those after the decimal point for π?

   b) Which will give the most accurate result for a circumference, using π ≈ 3.14 or using π ≈ $\frac{22}{7}$?

   c) Use π ≈ 3.14, then π ≈ $\frac{22}{7}$ to find the circumference of the Earth, given that the diameter is 12 766 km.

   d) Roughly, what is the difference in kilometres between the two results?

   e) Which result is more accurate?

6 a) Which one of these rules is correct for the circumference of a circle?

   (i) Circumference = 3 × Radius × π   (ii) Circumference = 2 × Radius × π. Explain why.

   b) The radius of a circle is 2.4 cm. Use your rule from a) to calculate the cirumference of the circle (use π ≈ 3.14).

   c) Use the rule, Circumference ≈ 3.14 × Diameter, to check your result in b).

CHALLENGE

7 Draw, as accurately as you can, a circle whose circumference is 16 cm. Inside your circle, write down its diameter as accurately as you can.

A8

ENRICHMENT

ACTIVITY

1   You need blank scrap paper and sticky tape.
     Here is how to make two cones.

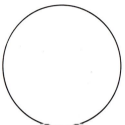

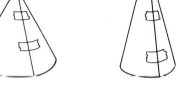

Draw a circle.   Cut into two sectors.   Each sector will make a cone.

a) Draw a circle with diameter 8 cm and cut it out.

   Use it to make two identical cones.

b) What is the circumference
   of the circle?
   (Use $\pi \approx 3.14$.)

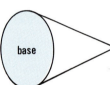

base

8 m

c) What is the circumference
   of the base of
   each of your cones?

d) Draw another circle with diameter 8 cm.
   Use it to make two more cones.
   This time the circumference of the base of
   one should be three times the circumference
   of the other.

   On each of your cones write down the
   circumference of its base.

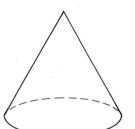

ASSIGNMENT

2   You need thin card or stiff paper, scissors, and sticky tape.
     Make this lampshade from thin card or stiff paper.

     Explain how you managed to get the measurements
     correct for your shade.

16 cm

20 cm

24 cm

A8

REVIEW

An angle is an amount of turn.
Angles are measured in degrees.

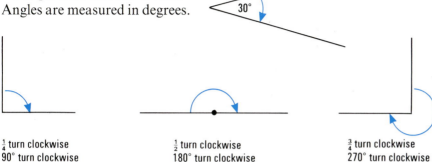

$\frac{1}{4}$ turn clockwise
90° turn clockwise

$\frac{1}{2}$ turn clockwise
180° turn clockwise

$\frac{3}{4}$ turn clockwise
270° turn clockwise

1 full turn clockwise
360° turn clockwise

We name the angle in the diagram using $\angle$ or $\frown$.

We write $\angle\,\mathrm{ABC} = x°$

or $\widehat{\mathrm{ABC}} = x°$.

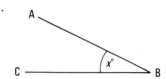

■ Draw an angle of   a) 280°   b) 190°.

CORE

CHALLENGE

1 You need tracing paper.

a) On your tracing paper draw two lines at an angle of about 60°.

b) Your two lines represent walls. Imagine that you are standing between the walls, at the point X. You want to walk away from X so that your path cuts the angle between the walls, in half. Draw the path you must take.

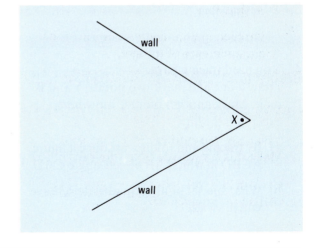

A9

2   One of the lines PA, QA, RA, SA, TA, cuts the angle CAB in half. Use your protractor to decide which one it is.

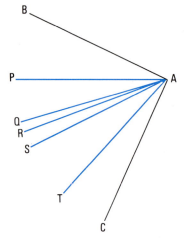

3   You need isometric dotted paper.
Draw the angle PQR.
Draw a line QX which cuts angle PQR in half.

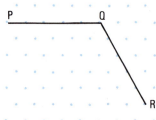

TAKE NOTE

The line which cuts an angle in half is called the *angle bisector*. We say that **BX** *bisects* angle ABC.

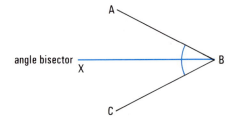

ACTIVITY

4   You need tracing paper.
Draw a line and mark two points A and B on it. Mark another point C anywhere, except on the line.

   a) By tracing and folding produce a figure ACBD in which AB bisects angle CAD.

   b) What special kind of quadrilateral have you drawn?

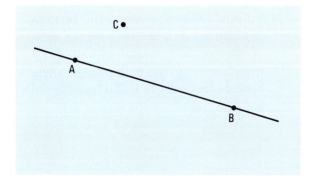

A9

5   PQRS is a quadrilateral.
    You can see only a part of it here.
    The dotted line bisects the
    angle at P and passes
    through R.
    Sketch and name all
    four special types of
    quadrilateral which PQRS
    can be.

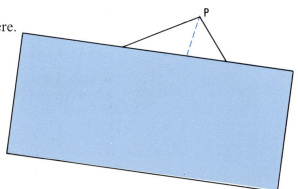

━━━━ ACTIVITY ━━━━

6   You need tracing paper.
    Draw any triangle ABC.

    a) By folding draw the three angle bisectors.
        Write down what you notice.

    b) Draw a circle inside the triangle which
        touches each side.

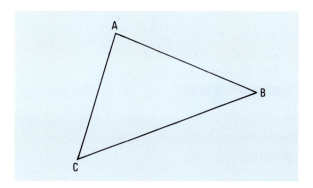

Perpendicular bisector of a line

━━━━ CHALLENGE ━━━━

1   You need tracing paper.
    On your tracing paper mark two points A
    and B approximately in the positions shown.

    An ant walks from the bottom edge of the
    tracing paper to the top edge, so that its
    distance from A is always equal to its
    distance from B. Draw the ant's path.

A•

•B

A9

2   One of the lines PQ, RS, TU, VW

  ● cuts XY in half, and also
  ● meets XY at right angles.

Which line is it?

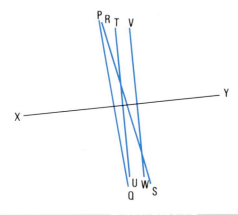

3   You need dotted squared paper.

Copy lines AB and PQ.
Draw lines which:

  ● meet AB and PQ at right angles, and also
  ● cut AB and PQ in half.

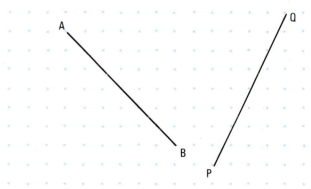

TAKE NOTE

The line which cuts another line in half and meets it at 90° is called the *perpendicular bisector* of the line.

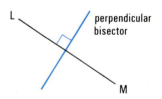

---

4   PQRS is a quadrilateral, but you can only see PQ.
The dotted line is the perpendicular bisector of PQ, and also of RS.
Sketch and name all the three special types of quadrilateral which PQRS might be.

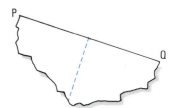

ACTIVITIES
---

5   You need tracing paper.

Draw any line AB on your tracing paper. By tracing and folding produce a rhombus with AB as its diagonal.

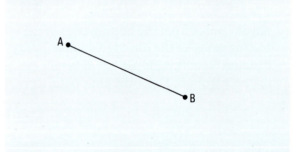

6   You need tracing paper.

a)   Draw any triangle ABC on tracing paper. By folding, draw the perpendicular bisectors of each side. Write down what you notice.

b)   Draw a circle around your triangle, to pass through A,B and C.

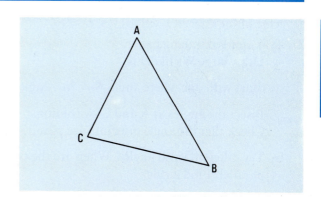

A9

ENRICHMENT

CHALLENGE

1   You need tracing paper.

These shapes can be produced by folding tracing paper (and without measuring any angles or measuring any sides).
Produce an example of each one yourself.

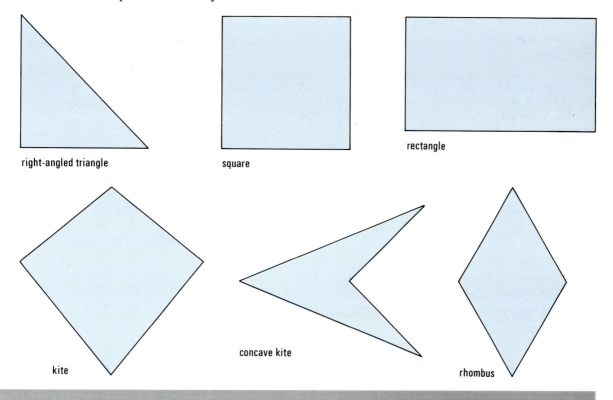

right-angled triangle   square   rectangle

kite   concave kite   rhombus

━━━━━━━━━━━━━━━━━━ EXPLORATION ━━━━━━━━━━━━━━━━━━

2 a) You need tracing paper.
       Draw large Ws like this.

       Start with one where angle A = 70°, angle B = 80° and angle C = 90°.

       Bisect the angles at A and C by folding.
       Check that the angle bisectors are parallel.

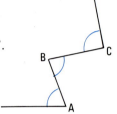

   b) Try different sized angles. When are the two bisectors
       parallel? Find a rule.

3   Think about quadrilaterals such as ABCD.
    For some quadrilaterals the angle bisectors
    at A and C are parallel. Which ones?

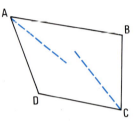

4   Draw a parallelogram.
    Draw the angle bisectors of a pair of
    adjacent angles as shown. What can you say
    about the angle at which they meet?

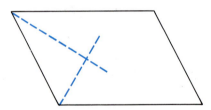

5 a) Draw a 40° angle.   Draw another line to make   Make these exterior angle
                                  a triangle.   bisectors by folding.

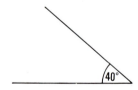   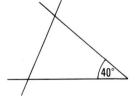   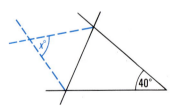

                                                                   Measure the angle marked x°.

   b) Repeat the experiment, producing a
       different triangle each time (keep the
       starting angle at 40°). What do you notice
       about the angle marked $x°$?

Electricity

1 Electricity is sold by the *unit*.

The illustrations show approximately how many units various electrical appliances use.

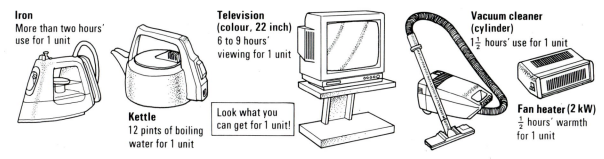

Iron
More than two hours' use for 1 unit

Kettle
12 pints of boiling water for 1 unit

Look what you can get for 1 unit!

Television (colour, 22 inch)
6 to 9 hours' viewing for 1 unit

Vacuum cleaner (cylinder)
$1\frac{1}{2}$ hours' use for 1 unit

Fan heater (2 kW)
$\frac{1}{2}$ hours' warmth for 1 unit

a) Roughly, how many units do you use in:

(i) vacuum cleaning the house for $\frac{1}{2}$ hour
(ii) boiling 24 pints of water?

b) Estimate how many units you would use to iron a shirt.

2 When you pay your electricity bill you pay for the number of units you have used, and for the hire of the meter (fixed charge).

From the bill find:

a) how many units were used by Mrs Maddox during the third quarter of the year

b) how much she paid for each unit she used

c) how much she paid in VAT

d) the total amount she paid.

Mrs L Maddox
Flat 1B
Cardigan Mansions
Worley St
Cokeford

THIRD QUARTER PAYMENT

METER READING DATE   9 / 10 / 1989

DATE OF ISSUE   11 / 10 / 1989

TARIFFS AND METER READINGS

PRESENT	PREVIOUS	UNITS SUPPLIED	FIXED CHARGES	AMOUNT	VAT%
70286	69857		6.35		0.00
		UNITS 5.850p		25.10	

AMOUNT NOW DUE AND PAYABLE

Gas

1   Gas is sold by the *therm*.

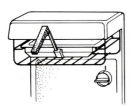 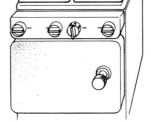

7 hours on full from 1 therm   9 hours from 1 therm   26 hours on Mark 5 from 1 therm

Estimate how many therms you would use to:

a) grill your breakfast toast every Saturday for 1 year
b) boil 5 kg of potatoes.

2   When you pay your gas bill you pay for the number of therms you use. Your gas bill also tells you the 'calorific value' of the gas you are using, that is, how much heat you get from the gas you use (measured in MJ/m^3, that is, megajoules per cubic metre; or Btu/ft^3, that is, British thermal units per cubic foot).

From Mr O'Donnell's gas bill find:

a) how many therms Mr O'Donnell used during the fourth quarter

b) what the meter reading measures (it is not the number of therms)

c) how the number of therms used are calculated

d) the total amount to be paid.

Mr K O'Donnell Flat 1A Cardigan Mansions Worley St Cokeford	4th QUARTER PAYMENT	DATE 08 / 1 / 89
		CALORIFIC VALUE 37.90 MJ/m^3
		(1016 Btu's per cubic foot)

DATE OF READING	METER READING		GAS SUPPLIED		VAT %	CHARGES £
	PRESENT	PREVIOUS	CUBIC FEET (HUNDREDS)	THERMS		
04 Jan	2361	2289	72	75.024	0.00	
				STANDING CHARGE		9.20
CREDIT TARIFF				38.50 Pence per Therm		28.88
				TOTAL AMOUNT DUE		

Telephone

1   You pay British Telecom for the number of units you use. For local calls 1 unit will give you between about 1 minute and 6 minutes of telephone time depending upon the time of day. Long distance and international calls cost more for 1 minute (you use more units in 1 minute).

From Mr Maddox's telephone bill find:

a) the number of units he used between 15 May and 18 August

b) the total cost of the units

c) the amount he paid in VAT

d) the total amount payable.

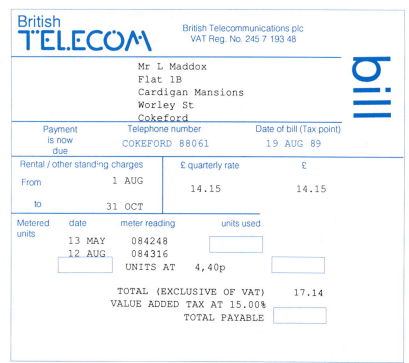

2   The table shows the cost of telephoning from kiosks for local calls and calls up to 35 miles away, at different charge rates. The pie chart shows when the charge rates apply.

a) How much would it cost to make a local call of ten minutes duration:

(i) at 5:00 p.m.   (ii) at 11:00 a.m.?

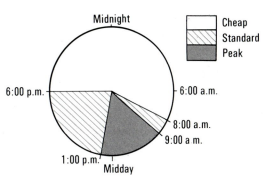

		Time for 1 unit (seconds)
Local	Cheap	360
	Standard	90
	Peak	60
National (up to 35 miles) (a)	Cheap	100
	Standard	34.3
	Peak	25.7
	1 unit costs 10p (inc. VAT)	

b) You want to telephone a friend 30 miles away. You have £1.

How many full minutes can your call last:

(i) if you phone at 6:30 a.m.

(ii) if you phone at 8:30 a.m.

(iii) if you phone at 10:30 a.m.?

ASSIGNMENT

3   Imagine you are moving into your own flat.
Collect some examples of electricity, gas and telephone bills (or use the ones in this chapter).
Work out how much you think it would cost you *each quarter* for gas, electricity and the
telephone.

Explain how the gas and electricity would be used. You will need to think about the kinds of
appliances you would want. You may also need to find out how much electricity or gas you would
expect to use on heating, lighting, etc. (Don't forget rental charges for meters, telephones, etc.)
Write out a full account of your costs and calculate the full cost for the year.

REVIEW

We often use charts, tables, graphs, etc., to help us in our thinking and to represent ideas more simply.

- A line graph, for example, can show a trend:

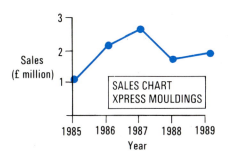

- A pictogram shows statistical data in an interesting and attractive way:

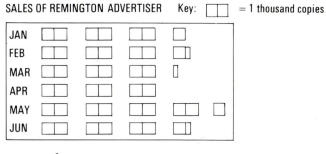

- A mileage chart gives a great deal of information in a compact space:

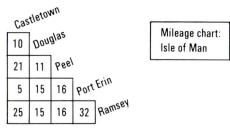

- A flow diagram helps to show us the order in which events take place, or gives us instructions:

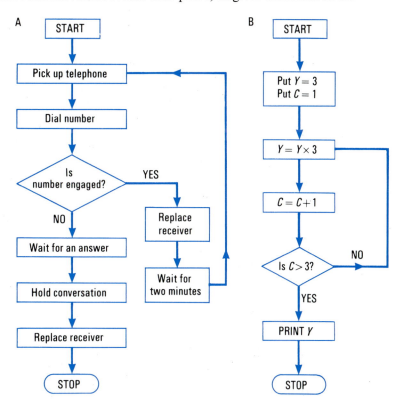

- Approximately, how many pounds worth of **XPRESS** mouldings were sold in 1987?

- How many miles is it from Peel to Port Erin?

- Approximately, how many copies of the Remington Advertiser were sold in March?

- What does the telephone flow diagram (A) tell you to do if the number is engaged?

- In flow diagram B,
  $Y = Y \times 3$ means
  'Y becomes $Y \times 3$'.
  $C = C + 1$ means
  'C becomes $C + 1$'.
  What number is printed out?

Reading and drawing charts

1   The temperature in Trafalgar Square was
    taken at hourly intervals from 10:00 to 15:00.
    The graph represents the results.

    a) What was the temperature at 11:00?

    b) What was the difference in temperature
        between 12:00 and 15:00?

    c) The graph suggests that the temperature
        at 13:30 was 10 °C. Why is this not
        necessarily true?

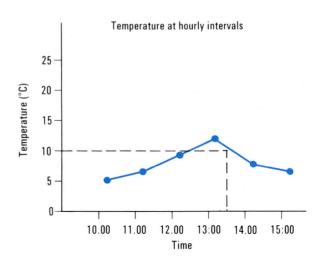

2   You need 1 mm squared paper.
    The table shows the times (in minutes and seconds) for the twenty members of Bath City Joggers'
    Club to cover the first kilometre in the London Marathon race. The figures in brackets give their ages.

JN (24)	4:28
PD (22)	4:30
TK (50)	6:44
NM (59)	7:32
MP (33)	4:01
TT (31)	4:55
SN (42)	5:30
ZB (44)	6:07
NK (29)	4:37
TL (23)	6:40
PP (38)	6:16
KL (32)	5:54
MD (54)	7:01
TC (49)	7:14
LG (66)	9:20
FT (41)	8:32
MB (25)	4:13
KV (30)	4:48
LC (39)	4:17
VV (20)	4:58

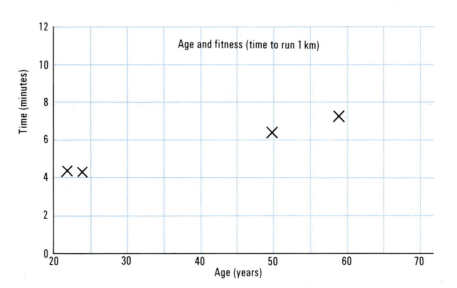

The times and ages of the first four runners have been marked on the scatter graph.

a) Copy the scatter graph. Mark the times and ages for all the runners.

b) Write one or two sentences to explain what your graph suggests about people's ages and fitness.

c) From your graph find a person who appears to be
    (i) very fit for his or her age group   (ii) very unfit for his or her age group.

3   Choose four towns or villages close to you. Write down the approximate distance, in kilometres, of each town from your own town or village.

Design a distance chart like the mileage chart in the *Review* to show the approximate distances between the towns or villages you choose.

ACTIVITY

4   You need squared paper.
The diagram helps you to sort four-sided shapes (quadrilaterals) into four different sets A, B, C and D.

a) On squared paper draw a parallelogram. Into which set, A, B, C or D, is this sorted by the diagram?

b) Draw examples of shapes which would be sorted into sets B, C and D.

c) True or false: all the shapes which are sorted into set A have either line symmetry or rotational symmetry? If you say 'false', draw a shape to explain your answer.

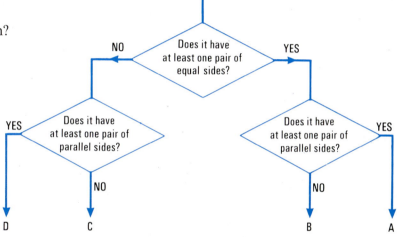

A11

5   Draw your own diagram to sort things into four different sets (for example, to sort vegetables, or fruit, or dogs,…). For each set name one of the members (a type of vegetable, fruit, dog, etc.).

ACTIVITY

6   Imagine you work in a pizza parlour, where pizzas can be eaten on the premises. You have a holiday job taking orders and serving customers at their tables.
Pizzas come in three sizes: small, medium and large. The parlour serves vegetarian and meat pizzas, and there are different types of each. Design an order form which you could use to take orders from your customers.
Remember your customers might want other things with their pizzas - drinks, salads, and so on. You will need to write down prices and the total cost on your order form.

7 a) Follow through the flow chart using the numbers in the list:
What does the flow chart do?

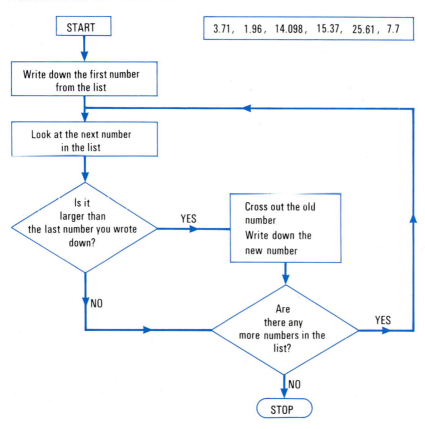

START

3.71, 1.96, 14.098, 15.37, 25.61, 7.7

Write down the first number from the list

Look at the next number in the list

Is it larger than the last number you wrote down? —YES→ Cross out the old number
Write down the new number

NO

Are there any more numbers in the list? —YES→

NO

STOP

b) Draw a flow chart for finding the *mean* of the numbers in a list.

CHALLENGE

8 This flow chart is for a computer program.

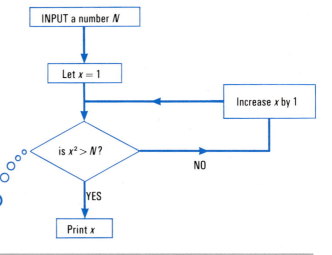

a) Choose N to be 30.
Work through the chart. Check that the number which will be printed is 6.

b) What number is printed when N is chosen to be 159?

c) Write one or two sentences to explain what the flow chart does.

is $x \times x$ larger than N?

INPUT a number N

Let $x = 1$

Increase x by 1

is $x^2 > N$?

NO

YES

Print x

9   The table shows the results of a survey about
    the employment of women in a town.

	Employed women	Unemployed women
Married women over 21	1217	875
Single women over 21	487	106

a) How many married women over 21 are
   not employed?

b) How many single women over 21 have
   jobs?

c) The information was collected using a questionnaire (a sheet with questions to be answered).
   The questions could all be answered by putting ticks in boxes, for example,

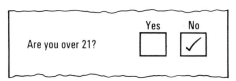

Design the questionnaire yourself.
Make sure you ask all the questions which are needed to enable you to complete a table like
that above.

A11

ASSIGNMENT

10 Design a survey of your own which needs a questionnaire. Choose your survey so that you can
    present the results in a table like that in question 9.

You can choose any kind of survey you wish. When you carry out your survey, ask at least 20
people to answer the questions on your questionnaire.

Write a report which explains what you did, and which shows your results.

CORE

Interest

1   When you put money into a savings account – say in a building society – you receive *interest*.

For example, at the time the advertisement was produced the Surfleet Moneybox savings account offered interest of 7.75% per annum (per year).

Everything you pay in earns 7.75%*
Why leave cash lying around?
Put it into a Surfleet moneybox
and earn good interest on it
*Interest rate is variable

a) What is 1% of £1000?

b) What is 7% of £1000?

c) What is 7.75% of £1000?

d) Which of these gives the correct interest you receive on £1000 after 1 year?

A   C . 7 7 5 × 1 0 0 0 =

B   C . 0 7 7 5 × 1 0 0 0 =

C   C 7 . 7 5 × 1 0 0 0 =

e) How much do you have in your Surfleet Moneybox savings account after 1 year if your initial saving is: (i) £1000   (ii) £500   (iii) £5000?

f) Explain why this will give the total amount in your savings account after 1 year if your initial saving is £800:   C 1 . 0 7 7 5 × 8 0 0 =

g) Use the method in f) to find the total amount after 1 year when the initial saving is £1750.

WITH A FRIEND

2   Imagine you invest £2000 in the Surfleet Moneybox savings account and you leave the money in the account for two years.

a) Each of you find how much will be in the account at the end of two years. Compare your results.

b) Here are the results obtained by two other pupils: Pat   £2310   Lesley   £2322.01
Either could be correct.

Decide between you how the two results were calculated, and write down an explanation for each. Which method of calculation do you think building societies use, Pat's or Lesley's?

▰▰▰▰▰▰▰▰▰▰▰▰▰▰ TAKE NOTE ▰▰▰▰▰▰▰▰▰▰▰▰▰▰

There are two main methods of calculating interest.
Suppose you have £1000 invested for 2 years at 9% interest.

Method A

After 1 year, interest paid	=	£90
After 2 years, interest paid	=	£90 × 2
Total interest	=	£180

This is called *simple interest*.

Method B

After 1 year, interest paid	=	£90
Total in the account	=	£1090
Interest for year 2	=	£1090 × 0.09
	=	£98.10
Total interest = £90 + £98.10	=	£188.10

This is called *compound interest*.

3   When you borrow money, or 'take out a loan', you have to pay interest to the lender. You also pay interest on overdrafts at the bank and on money unpaid on credit card accounts. The interest you pay can vary a great deal. You should always check how much you will be paying before you take out a loan.

a)   The table shows how much you pay back each month on different sized loans, when you take out a loan with a company called 'Moneymate'.
Check that if you borrow £2000 for 60 months you pay £52.77 per month.

MONEYMATE
FAST PERSONAL LOANS!
TYPICAL MONTHLY REPAYMENTS

AMOUNT	36 MTHS	60 MTHS	120 MTHS	180 MTHS
£800	30.39	21.92	———	———
£2000	74.13	52.77	38.39	36.36
£3500	122.39	84.92	57.83	63.63
£7500	275.41	195.09	140.64	129.36

b)   How much do you pay each month if you borrow £3500 and pay the money back over a 3 year period?

c)   Calculate the total amount you pay back when you borrow:

(i)   £800 over a 5 year period
(ii) £7500 over a 3 year period.

d)   (i)   How much do you pay in interest if you borrow £7500 and pay it back over a 5 year period?

(ii) On average, how much interest is this per year?

(iii) Use your result in d) (ii) to check that this represents a rate of simple interest of about 11.2% per year on £7500.

e)   What rate of simple interest do the repayments on £2000 borrowed for 3 years represent?

Credit

1 a) When you buy an expensive article from a
       shop – say a television – you can either

       • pay cash
   or • pay in instalments (credit sale).

   This television costs £452
   if you pay cash.
   How much do you pay
   if you pay by instalments?

CASH PRICE £452

Credit sale
Deposit: £50
then 10 monthly
instalments
of £45

A12

   b) When you pay by instalments, you are really borrowing money from the store, and paying
       back over a period of time. For this service the store will charge you interest. That is why the
       cash price and credit sale price of the television are different.
       How much more do you pay for the television if you choose credit sale instead of paying cash?

   c) What is this difference in price as a percentage of the cash price?

2 The table shows the credit terms for buying second-hand cars from Blackbird Motors.

BLACKBIRD MOTORS

Cash price (£)	Deposit (£)	Amount borrowed (£)	Repayment on amount borrowed		
			Credit period		
			1 year	2 years	3 years
625	125	500	555	610	665
1250	250	1000	1110	1220	1330
2500	500	2000	2220	2440	2660
3750	750	3000	3330	3660	3990
5000	1000	4000	4440	4880	5320
6250	1250	5000	5550	6100	6650

   a) Why do you think you have to pay more if the payments are over 3 years rather than 2 years or
       1 year?

   b) Imagine you want a car whose cash price is £6250.
       The total you pay on credit sale if you pay over 3 years is £1250 + £6650.
       How much more is this than the cash price?

   c) What is the difference between cash price and credit sale price, as a percentage of the cash price?

   d) How much more would you pay on credit sale than cash for a £2500 car bought over 3 years?

Hire purchase

1 Credit terms are sometimes also called 'Hire Purchase' terms (HP terms). Discuss between you why buying this way is sometimes called '*Hire* Purchase'. Write down what you decide.

2 a) Copy the HP calculation card and use it to find the monthly instalments on the motor bike.

	£
Cash price	800
Deposit (30%)	
Balance left to pay	
Interest to be paid on Balance (20%)	
Total repayments over 12 months (Interest and Balance)	
Monthly repayments	

MOTOR BIKE 100cc
Maximum speed 120 km / h
Tank capacity 7.8 litres

• Cash price £800
• HP terms available
Deposit 30%
12 monthly instalments. Interest rate 20%

b) How much more do you pay for the motor bike on HP rather than using cash?

3 The interest rate for Hire Purchase and Credit schemes is usually called the *annual percentage rate* (APR).
The table shows the repayment you make each month on each £100 you borrow at different APRs, over different loan periods.

You buy a music centre for £400 on HP, and pay back over 3 years. The APR is 22%.

a) How much do you pay per month?

b) How much do you pay altogether over the 3 year period?

c) How much more do you pay by this method than by paying cash?

Loan period		Monthly repayment				
		1 year	2 years	3 years	4 years	5 years
APR	20%	£9.19	£5.01	£3.63	£2.96	£2.56
	22%	£9.27	£5.09	£3.72	£3.05	£2.65
	24%	£9.35	£5.17	£3.80	£3.13	£2.75
	26%	£9.50	£5.33	£3.97	£3.31	£2.93

A12

4 Investigate different ways of investing £1000 for 2 years.
   Find the interest rates which are offered, and find out how the interest is calculated.
   Write a report which compares the different amounts of interest you can receive from different
   forms of investment (such as bank accounts, building societies, shares, unit trust funds).
   Say which you think is the best investment and why.

A12

REVIEW

- The larger shape is an *enlargement* of the smaller shape.

  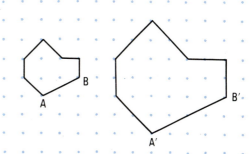

  All lengths on the larger shape are twice the corresponding lengths on the smaller shape.

  For example, $A'B' = 2 \times AB$.

  The corresponding angles of each shape are equal. We say that the two shapes are *similar*.

■ Sketch two quadrilaterals which have the same angles, but which are *not* similar.

- Triangles which have the same angles are *always* similar. This is not true for other shapes.

- In similar shapes the ratios of corresponding sides are equal.

  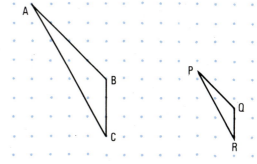

  For example, $\dfrac{AB}{BC} = \dfrac{PQ}{QR}$.

  Also, because

  ABC is an enlargement of PQR, $\dfrac{AB}{PQ} = \dfrac{BC}{QR}$.

■ *Sketch* two similar triangles, and call them LMN and RST. Write down as many sets of equal ratios as you can.

CONSOLIDATION

WITH A FRIEND

1 a) Decide between you which of these are examples of similar shapes or solids. Write down what you decide.

     (i) Your thumbprint now, and your thumbprint when you were a baby.
    (ii) A cat and its kitten.
   (iii) A car and a model of the car.
   (iv) A picture in a newspaper and a reduced photocopy of it.
    (v) The pattern on this stamp and the pattern it makes on paper.
   (vi) The Post Office tower and a 2D drawing of it.

  b) Think of two examples of your own of pairs of similar shapes.

ACTIVITY

2   You need isometric dotted paper.
    The black lines are part of a large
    shape.
    The blue lines are part of a small
    shape.
    The two shapes are similar,
    and AB is an enlargement of PQ.
    Copy and complete both shapes.

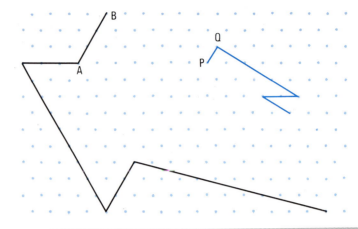

3   a) One of these sets contains three
        similar shapes. Which one?

    b) Explain why the shapes in each of
        the other two sets are not similar.

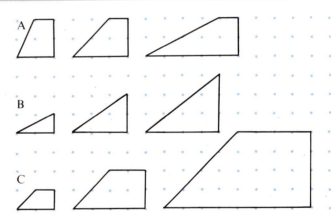

4   Two of the triangles A to E
    are similar. Which two?

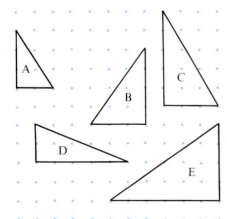

5   You need 1 cm squared
    dotted paper.

    a) Draw a shape which is similar to this:

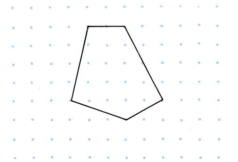

    b) Draw a shape which has the same
        sized angles, but which is *not* similar
        to the shape in a).

6   Do not use a protractor or
    ruler. The shapes are not
    drawn to scale.

    The two shapes are similar.
    One is an enlargement of
    the other. A corresponds to
    E, and B corresponds to F.

    a) How many degrees is
        the angle marked

        (i) ▲   (ii) ●?

    b) How long is

        (i) AB   (ii) FG?

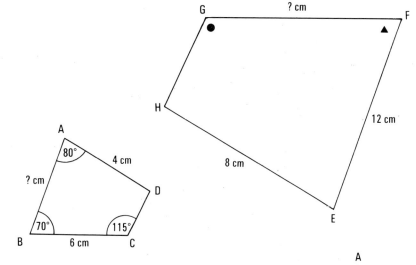

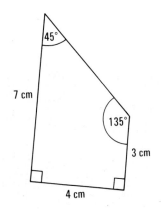

7   The two shapes are similar.

    a) *Sketch* the larger shape.

    b) Mark in the sizes of the
        angles of the larger
        shape on your sketch.

    c) Mark in the lengths of
        AD and BC on your
        sketch.

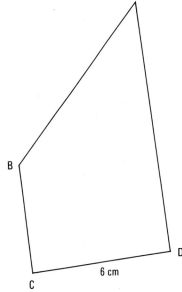

A13

8   The shadow of the larger
    tree is 18 m long.
    How long is the shadow of
    the smaller tree?
    Explain how you arrived at
    your result.

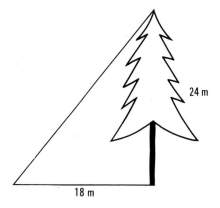

ENRICHMENT

1 a) Collect an example of a pair of bottles, boxes or other containers which are similar.

   Write a short report explaining how you know that they are similar. If you cannot find an example, choose two objects which *appear* to be similar but which are not. Write a report to explain why they are not similar.

   b) Make two cones which are similar. Write a short report explaining how you know that they are similar.

2 a) Draw any triangle ABC.
   Imagine that you keep increasing the lengths of the sides by 1 cm. Predict what will happen to the ratio $\frac{AB}{BC}$ and to the size of $\angle ABC$.

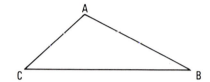

   b) Draw the first three larger triangles. Check your predictions in a). Write about what you discovered and modify your predictions if necessary.

3 You want to find the height of an industrial chimney. It is a sunny day. Explain how you can calculate the height by measuring shadows.

REVIEW

- Parallel lines can be
  drawn using a ruler
  and set square.

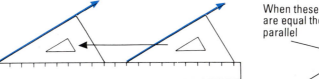

When these angles
are equal the lines are
parallel

The diagrams show the pairs of equal angles formed when a line (transversal) is drawn across two parallel lines.

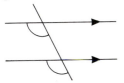 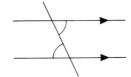 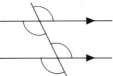 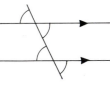

TAKE NOTE

Pairs of equal angles formed when a line (transversal) is drawn across two parallel lines are given these special names:

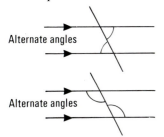

Alternate angles

Alternate angles

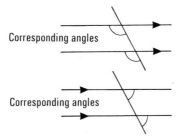

Corresponding angles

Corresponding angles

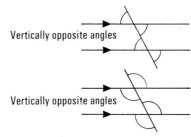

Vertically opposite angles

Vertically opposite angles

CORE

ACTIVITY

1 a) In the first diagram there are three pairs of parallel lines.
      Sketch the diagram. On your sketch mark with an
      a all those angles equal to the angle marked *a*.
      (There are seven others, for a total of eight.)
      Do the same for the angles marked *b* and *c*.

   b) Use a large sheet of plain paper, a ruler
      and set square. By drawing as many lines
      as you wish, produce a seven-pointed star
      which has exactly three pairs of equal 'point-angles'.

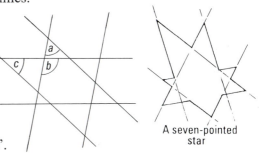

A seven-pointed
star

B1

2   This is how to make a 'Moonstar' shape.

Draw three sets of parallel lines, crossing anywhere. Here are three examples:

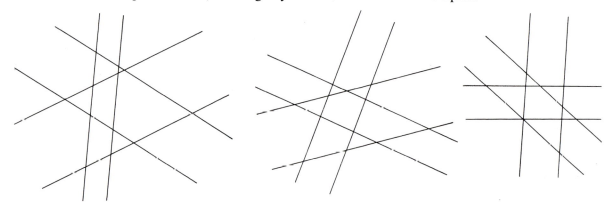

Your three pairs of parallel lines will define a maximum area polygon (a 'Moonstar'). The three maximum area Moonstars for the sets of parallel lines above are:

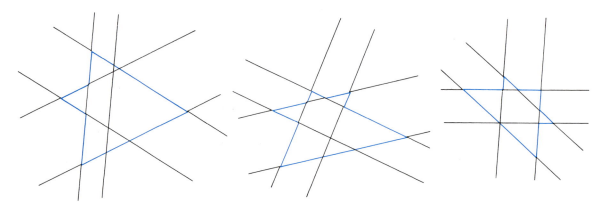

Investigate the different kinds of Moonstar shapes that can be produced. How many sides and points can they have? How many of the points have equal angles? What symmetries can Moonstars have? Write a report about what you discover.

This is a six-sided Moonstar with four points, two pairs of equal point-angles, and turn (rotational) symmetry of order 2.

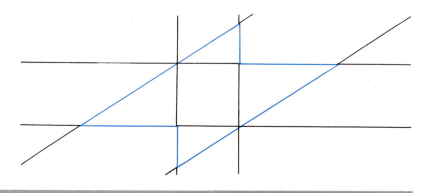

Angles of triangles and parallelograms

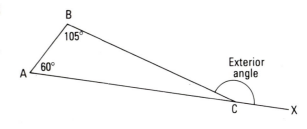

1   a) In the diagram, angle BCX (∠ BCX) is
called an *exterior angle* of the triangle. By
first working out the size of ∠ BCA, find
the size of ∠ BCX.

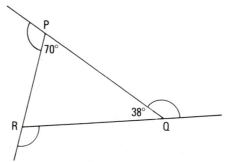

b) In the triangle PQR, all three exterior
angles have been drawn. Find the size of
each.

c) There is a relationship which connects
each exterior angle of a triangle with two
of the interior angles. Find it, and write
the relationship in your own words.

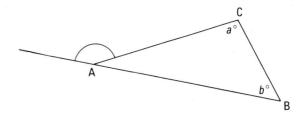

d) (i) Explain why ∠ CAB is
$180° - (a° + b°)$.
(ii) Explain why the exterior angle
measures $a° + b°$.

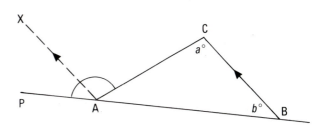

e) Use what you know about the angles
between parallel lines to explain why the
size of the exterior angle CAP is $a° + b°$.

════════════ TAKE NOTE ════════════

The exterior angle of a triangle equals the sum
of the two opposite interior angles.

$n° = a° + b°$

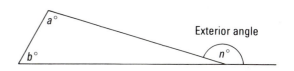

2   In each diagram find the
size of the angle marked '?'.

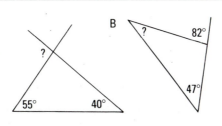

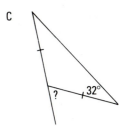

B1

3 a) Triangle A is given a
      half turn to make
      parallelogram B. Sketch
      the parallelogram; mark
      the size of each angle.

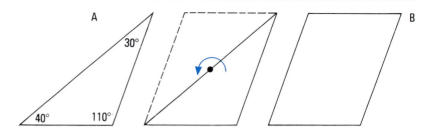

   b) Triangle A is again given a
      half turn to make
      parallelogram C.
      Sketch the parallelogram;
      mark the size of
      each angle.

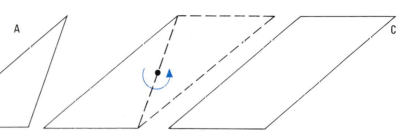

   c) Sketch the third parallelogram that can be made from triangle A; mark the size of each angle.

   d) Look at your three parallelograms; what can be said about

      (i) any pair of
           opposite angles

      (ii) any pair of
           adjacent angles?

CHALLENGE

   e) Look at the diagram.
      Use it to explain your
      answer to d) (i) and (ii).

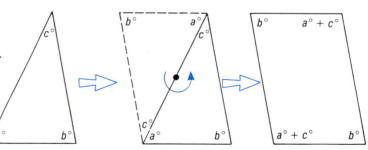

TAKE NOTE

Opposite angles of a parallelogram are equal.
Adjacent angles add up to 180°.

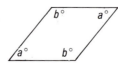
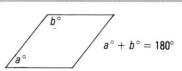

B1

THINK IT THROUGH

4 a) Sketch the diagram and find the size of
      each of the other angles, marking them
      on your sketch.

   b) Two of the lines in the diagram are
      actually parallel. Mark them with arrows.

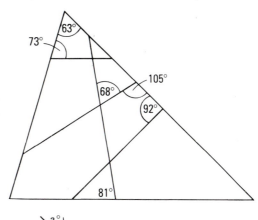

5 a) Explain why triangle ABC is isosceles.

   b) Find the size of each angle of the triangle
      if *a* is 37.

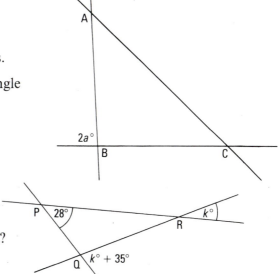

6 The drawing cannot be correct. Why not?

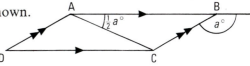

7 ABCD has not been drawn correctly.

   a) Draw it correctly, so that it agrees with the angles shown.

   b) What special kind of figure is the true
      shape ABCD?

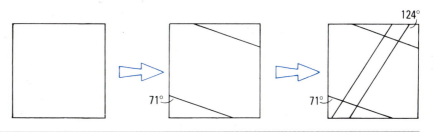

8 The square has two sets
   of parallel lines drawn
   inside it. Make a large
   sketch of the final
   diagram. Mark the sizes
   of all the angles on it.

ENRICHMENT

Parallel line challenge

1   In the diagram AE is parallel to BD. (We write AE ∥ BD.)

BC = $\frac{1}{3}$ AC, AE = 24 cm, and DC = 10 cm.

a) Explain why triangles ACE and BCD are similar.

b) Find the length of BD and EC.

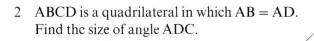

2   ABCD is a quadrilateral in which AB = AD. Find the size of angle ADC.

3   In the diagram there are three pairs of parallel lines. Find x.

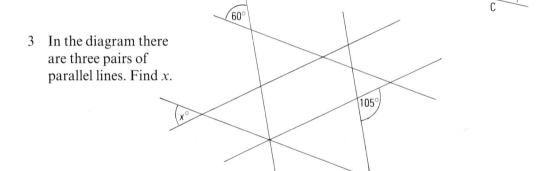

4   a) This is a piece of folded card standing on a table. It is folded along a line of symmetry, and EBAD is a parallelogram. When the card is opened out flat upon the table, how many degrees is angle ABC?

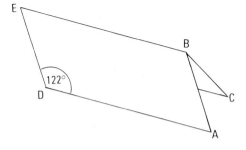

b) If ED = 10 cm, and AD = 20 cm, find the approximate area of the shape. Hint: you will have to use sin, cos, or tan, or a scale drawing, to find a distance.

B2 WORKING IN 2D AND 3D

CORE

B2

1 Here are some drawings of sheds. Draw the missing views.

3D view,	Elevation from X	Elevation from Y	Plan from Y
a) 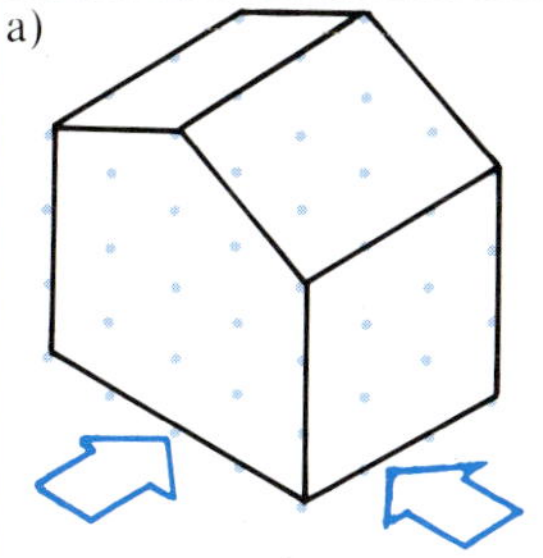	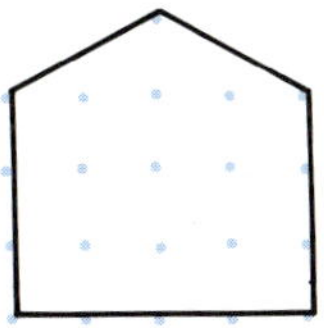		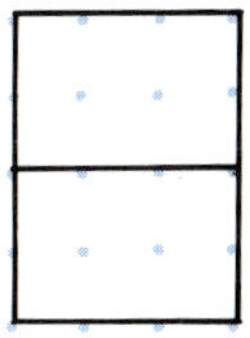
b) 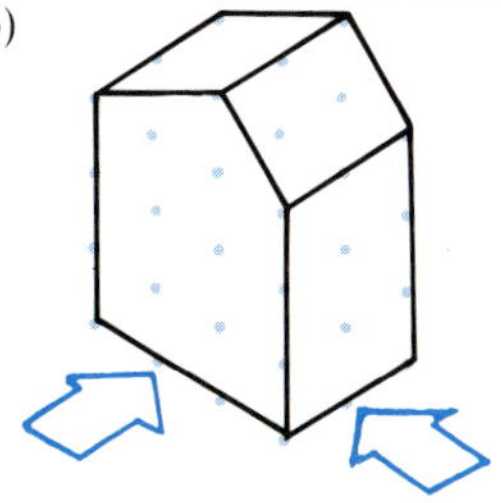		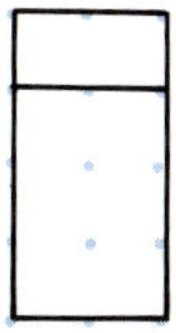	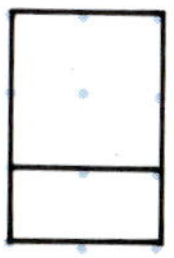
c) 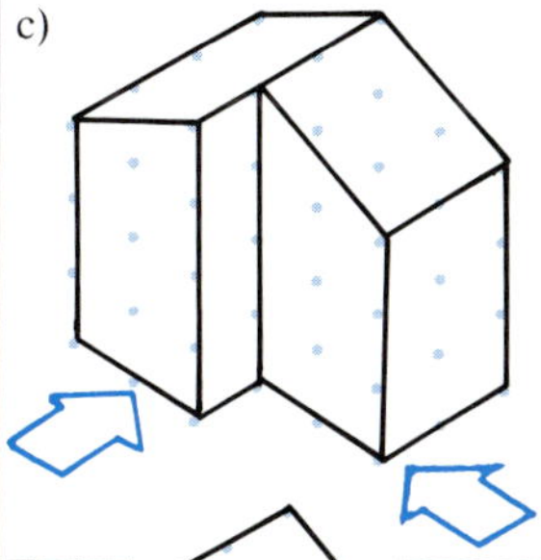		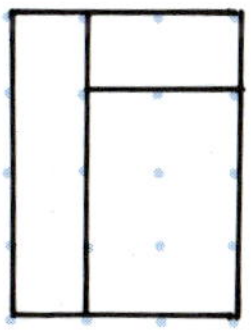	
d) 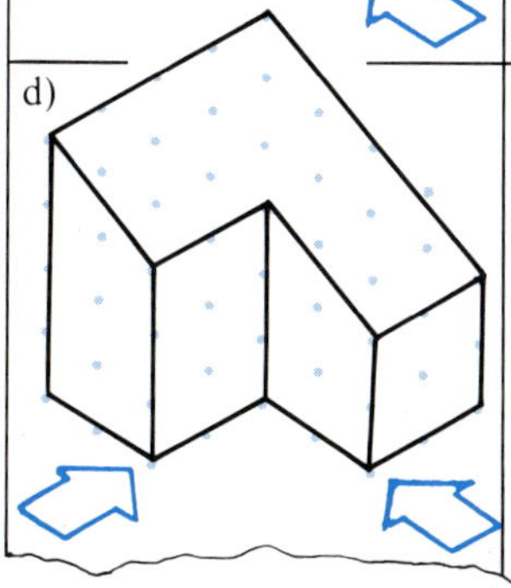		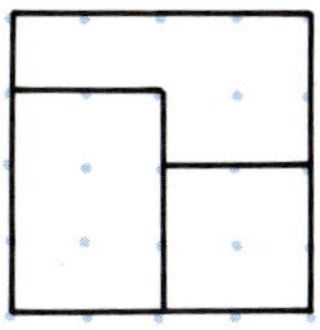	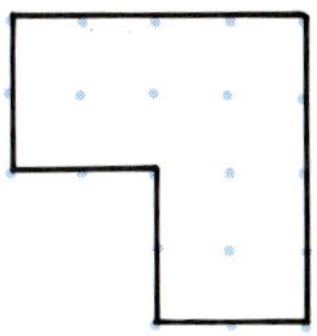

32

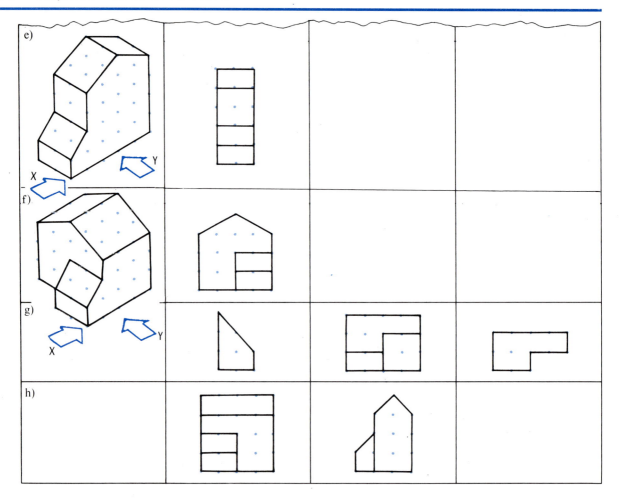

2   This is Horace's shed.
    Draw the elevations from
    X and Y.

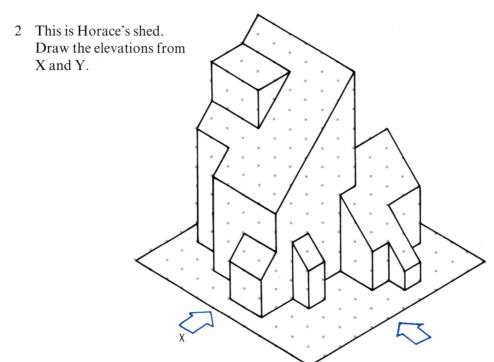

ENRICHMENT

1   A packet of cornflakes is
standing in the middle of a
round table marked like this:

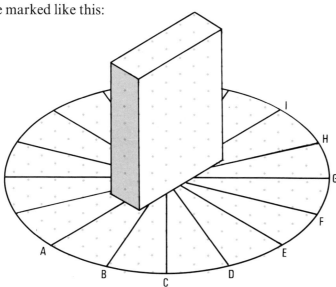

a) This elevation is   Is this elevation
    from the point C.   from B or from D?

   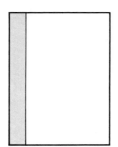

b) These three elevations are from F, G and H. Write down which is which.

(i) 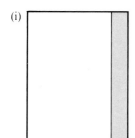   (ii) 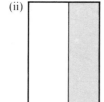   (iii)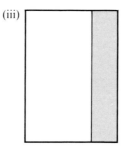

c) Sketch the elevation of the packet from B.

B2

2 The dot diagram is a drawing of a factory.

This is the plan of
the factory.

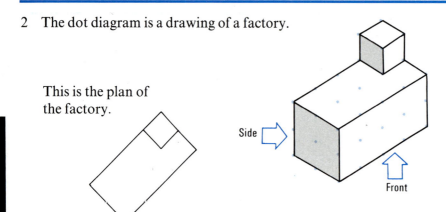

The plan can be used to draw an elevation, seen from the front.
Follow these steps:

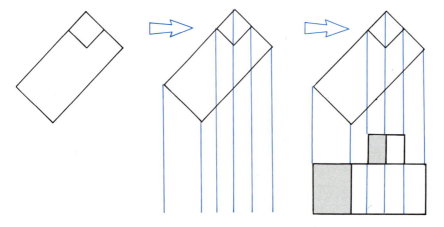

Use this method to draw the front elevation of this factory:

Start by drawing
this plan.

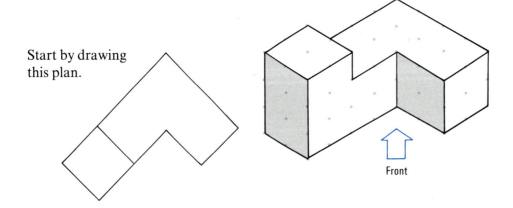

3   Here is a drawing of another factory.

Here is a plan
for the factory.

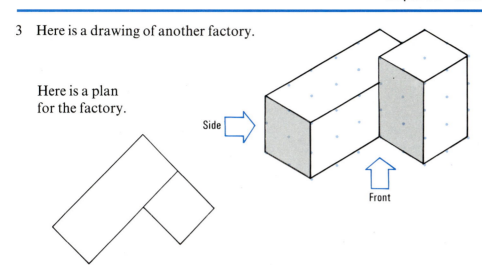

The plan can be used to produce a side *and* a front elevation, like this:

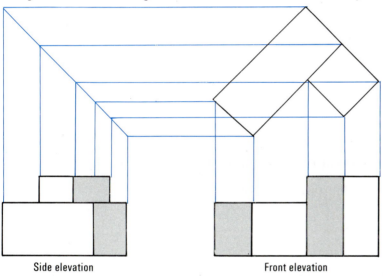

Side elevation   Front elevation

Use this method to draw the side and front elevations of this factory.

Start with this plan.

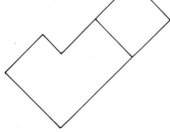

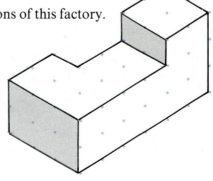

B2

B3 ESTIMATING

REVIEW

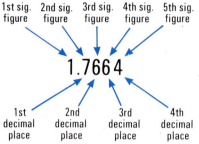

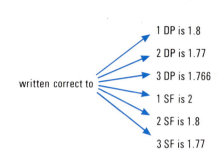

DP means 'decimal place'.
SF means 'significant figure'.

1.75
- correct to 1 DP is 1.8
- correct to 2 SF is 1.8

0.004 97
- correct to 1 DP is 0.0
- correct to 1 SF is 0.005

↑ 1st significant figure   The first significant figure in a number is the first digit which is not zero.

■ Which is the second significant figure in   (i) 10.84   (ii) 0.042 98?

■ Write each number correct to (i) 3 DP (ii) 4 SF: a) 17.004 29
                                                      b) 0.000 104 95.

CONSOLIDATION

1 a) Estimate the result of each of these by first rounding each number to 1 SF:

   (i) $6.76 + 19.845$ (ii) 8.79×3.46 (iii) $29.7 \div 1.98$.

   ↑   ↑
   $\approx 7 + \approx 20$

   b) Use your calculator to do the calculations in a). Check that your calculator results and your estimations roughly agree with each other.

2 A number written correct to 1 DP is 3.7.

   a) What is the *smallest* number which is written as 3.7 correct to 1 DP?
   b) What is the *largest* number which is written as 3.7 correct to 1 DP?

EXPLORATION

3 Investigate which is more accurate: ● writing a number correct to 3 DP
                                        or ● writing a number correct to 3 SF.

   Write down what you decide.

Estimations with whole numbers

1   The three results show three attempts at multiplying 123×497:

| _61131._ | | _611310._ | | _6113._ |

a) Write 123 and 497 correct to the nearest 100.

b) Use your approximations in a) to decide which result for 123×497 is most likely to be correct.

c) Use your calculator to check your choice in b).

2   The three calculator displays show three attempts at multiplying 568×263:

| _149384._ |
| _14938._ |
| _1493844._ |

a) Write 568 and 263 correct to the nearest 100.

b) Use your results in a) to decide which calculator result is most likely to be correct.

c) Do the calculation on your calculator.

━━━━━━━━━━━━━ TAKE NOTE ━━━━━━━━━━━━━

Whenever you do a calculation using your calculator *always* make a rough estimate first. Normally you will be able to make the estimate in your head.

3   a) In your head, decide which calculator display is most likely to be correct for the calculation $244 \div 48$.

| _50.8333333_ |
| _5.08333333_ |
| _508.333333_ |

   b) Do the calculation.
      Were you correct in a)?

4   a) In your head, decide which calculator display is correct for $12 \div 124$.

| _0.967741930_ |
| _0.096774193_ |
| _0.009677419_ |

   b) Do the calculation.
      Were you correct in a)?

B3

5   In your head, make an estimation for each calculation, then do the calculation. If your estimate and your calculator result do not appear to agree, check each of them.

a) 17×22

b) $\dfrac{76}{28}$

c) $\dfrac{15 + 50}{22}$

d) $\dfrac{150 - 77}{4}$

e) 186×301

f) $764 \div 26$

g) $17 \times 19 \times 12$

h) $\dfrac{67 \times 124}{307}$

WITH A FRIEND: THE ESTIMATING GAME

6   Each of you write down five calculations involving whole numbers, like those in question 5. Make them as complicated as you wish. Then each of you, in your head, estimate result for all ten calculations. Check your estimate with a calculator.

    Score: 1 point for the nearer estimate each time. The player with more points out of 10 wins.

B3

Estimations with decimals

1   The calculator displays show three attempts at finding the area of the advertisement:

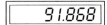

4.76 cm

1.93 cm

a) Write the length and width of the advertisement correct to 1 SF.

b) Use your results in a) to estimate the area of the advertisement.

c) Which of the three calculator results do you think is most likely to be correct?

d) Use your calculator to check your decision in c).

2   This is the formula for the radius (r cm) of a circle with circumference c cm: $r = \dfrac{c}{2\pi}$

a) In your head, estimate the radius of a circle with circumference

    (i) 6.2 cm   (ii) 17.9 cm.

b) Calculate the radius of each circle in a) correct to 2 SF (use $\pi \approx 3.142$). For each circle check that your estimate gives a reasonable idea of the radius. If it does not, check your estimate and your calculation.

3   The area of a triangle with base b cm and height h cm is $\frac{1}{2}bh$ cm^2.

a) Estimate the area when b is 17.5 and h is 12.8.

b) Calculate the area.

    Check that your estimate gives a reasonable idea of the radius.

4   The calculator displays show three attempts at the calculation

$$\frac{0.47 \times 3.71}{4.2 + 0.69}$$ | 0.356584867 | | 3.565848677 | | 0.035658486 |

  a) In your head, estimate the result, and decide which display is most likely to be correct.

  b) Do the calculation yourself.

5   Calculate each of these. Make sure you make a mental estimate for each one first.

  a) $\dfrac{31.7}{0.56} \times 0.48$ b) $\dfrac{7.61 + 19.8}{0.23 - 0.14}$ c) $\dfrac{365.6 - 47.9}{27.8 + 59}$ d) $\dfrac{\sqrt{96}}{0.4^2}$.

6   Write down five calculations like those in questions 4 and 5. Each of you, in your head, estimate a result for all ten calculations. Check your estimates with a calculator.
  Score: 1 point for the nearer estimate each time. The player with most points out of 10 wins.

B3

ENRICHMENT

1   The amount of money spent on Defence and Education in 1960 and 1983 is shown in the table (each figure correct to 2 SF):

	1960	1983
Defence	1600	16 000
Education	920	15 000
Expenditure in UK (£ million)		

  a) (i) Check that the maximum amount which might have been spent on Defence in 1960 is £1649 999 999.99.
      (ii) What is the maximum amount which might have been spent on Education in 1960?

b) (i) Check that the minimum amount which might have been spent on Defence in 1983 is £15 500 million.

   (ii) What is the minimum amount which might have been spent on Education in 1983?

c) (i) Check that the expenditure on Defence and Education together (£K) in 1960 is satisfied by the inequality:

   £2465million $\leqslant$ £K < £2575 million

   (ii) Write down a similar inequality for the expenditure on Defence and Education in 1983.

2   On an architect's drawing of the dealing room in the Lloyd's building the length and width are given as 59 m and 46 m correct to 2 SF.

a) What are the maximum length and width the dealing room might really have?

b) What are the minimum length and width the dealing room might have?

c) Complete the inequality:

   $\leqslant$ perimeter of room (m) <

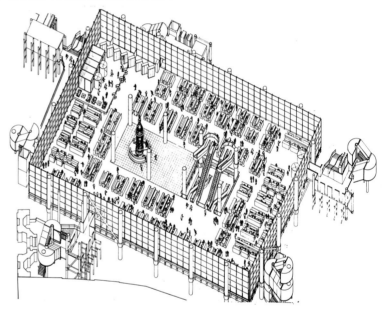

REVIEW

We have a shorthand way of writing multiplications:

$$2 \times 2 \times 2 \times 2 \times 2 \text{ is } 2^5$$

$$10 \times 10 \times 10 \times 10 \times 10 \times 10 \text{ is } 10^6 \text{ and so on.}$$

> The '5' and '6' are called indices.
> The '2' and '10' are called base numbers.

We say 'two to the power of five', 'ten to the power of six', etc.

Powers of ten have special names in our number system.

Number	Power of ten	Name
10	10^1	ten
100	10^2	one hundred
1000	10^3	one thousand
10 000	10^4	ten thousand
100 000	10^5	one hundred thousand
1 000 000	10^6	one million
1 000 000 000	10^9	one billion

■ Which is largest: 10^3, 3^{10} or 3×10?

CONSOLIDATION

1   a) *Guess* the approximate value of 2^{10}.   About 20
                                                    About 800
                                                    About 20 000?

    b) Check your guess with a calculator.

    c) Do the same for each of these: (i) 3^{10} (ii) 9^5 (iii) 20^6.

2   One of these graphs shows how the numbers $2^1, 2^2, 2^3, 2^4, \ldots$ grow for increasing powers of 2. Which one is it?

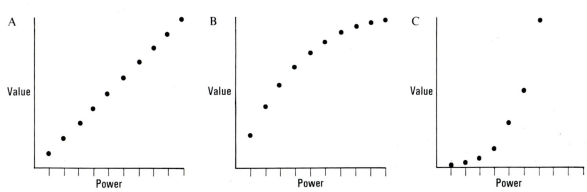

3 Write each of these as a power of 5:

a) 25   b) 625   c) 15 625.

4 Some numbers can only be represented using whole number powers in one way, for example,

$$7 = 7^1$$

Others can be represented in more than one way, for example,

$$8 \begin{cases} 2^1 \times 4^1 \\ 8^1 \\ 2^3 \end{cases} \qquad 12 \begin{cases} 12^1 \\ 2^2 \times 3^1 \\ 3^1 \times 4^1 \\ 2^1 \times 6^1 \end{cases} \qquad 36 \begin{cases} 36^1 \\ 6^2 \\ 2^2 \times 3^2 \end{cases}$$

Find all the numbers from 1 to 100 which can
only be represented using whole number powers in one way.

5 Write each of these as a single whole number:

a) 10^5 b) 2×10^5 c) 3^4 d) 2×3^4 e) $(2 \times 10)^5$.

B4

CORE

Negative indices

1 This table shows decreasing powers of 10. To
go from one line to the next we divide by 10.

a) Copy the table. Complete the entries
except for the final two lines.

b) What power of 10 would continue the
pattern for the final two lines?

 Stop here

Number	Power of ten	In words
1 000 000	10^6	one million
100 000	10^5	
10 000		ten thousand
1000		
	10^2	
	10^1	

We agree to let 10^0 mean 1 and to let 10^{-1} mean $\frac{1}{10}$ (or, 0.1).

c) Complete the last two lines of your table and add three more.

━━━━━━━━━━ THINK IT THROUGH ━━━━━━━━━━

d) Write as a power of 10:   (i) 0.01   (ii) 0.000 1   (iii) 0.000 000 01.

2   Copy and complete:

a) $5^? = \dfrac{1}{5}$
b) $10^? = \dfrac{1}{10^3}$
c) $2^? = \dfrac{1}{8}$
d) $10^{-6} = \dfrac{1}{10^?}$

e) $0.000\,1 = 10^?$
f) one millionth $= \dfrac{1}{10^?}$
g) $10^{-10} = 0.00\ldots$

3   Write as a power of 5:

a) twenty-five   b) one twenty-fifth   c) one fifth   d) one.

4   Write as a decimal: a) 2^{-1} b) 2^{-2} c) 5^{-1} d) 8^{-1}.

━━━━━━━━━━ TAKE NOTE ━━━━━━━━━━

We use positive powers to represent very large numbers:

10 000 000 000 000 000 000 000 000 km $= 10^{25}$ km.

We use negative powers to represent very small numbers:

0.000 000 000 000 001 mm $= 10^{-15}$ mm.

5   a) The mass of a virus is about 10^{-21} kg.
Write this (i) as a decimal (ii) as a
fraction.

b) How many viruses are needed to make a
mass of 10^{-20} kg?

c) A house spider weighs about 10^{-4} kg.

(i) How many grams is this?
(ii) Roughly, what is the weight in grams
of 100 house spiders?

B4

---

6   Find the missing numbers:

a) $8^? = 16\,777\,216$ b) $4^? = 0.015\,625$ c) $?^{-4} = \dfrac{1}{1296}$ d) $?^{-2} = 0.111\,111\,11\ldots$

---

7   a) On your calculator, press

C 3 x^y 2 = $\bullet$ $\circ$ $\circ$ $\circ$ $\circ$ $\bigcirc$ $\bigcirc$

You may need to use the inverse key, for example,

C 3 INV $\times$ $\overset{x^y}{}$ 2 = ...

and C 2 x^y 3 =

Write down each result. Try some more examples.

b) Try some negative numbers, for example,

C 3 x^y 2 +/− =

C 2 +/− x^y 3 =

Write down your results.

c) Explain what the x^y key on a calculator does.

B4

8   Use the x^y key on your calculator to find:

a) 3^{10}

b) 2.4^8 (to the nearest whole number)

c) 0.8^4

d) 9^{-2} (to the nearest 0.01)

e) $(-4)^{-2}$

f) $(-5)^{-1}$.

9   a) Calculate:

(i) 4^5 (ii) 5^{-2} (iii) $4^5 \times 5^{-2}$.

b) Calculate:

(i) 2.1^3 (ii) 0.3^{-2} (iii) $2.1^3 + 0.3^{-2}$.

c) Calculate:

(i) 0.8^3 (ii) 0.4^5 (iii) $\dfrac{0.8^3}{0.4^5}$.

Powers of 10

1   We can write 300 as 3×10^2. Copy and complete each of these:

a) $800 = 8 \times 10^?$ b) $2\,000\,000 = 2 \times 10^?$ c) $190\,000\,000 = 19 \times 10^?$.

2   Notice that $3 \div 10$ and 3×10^{-1} $\left(\text{and } 3 \times \dfrac{1}{10}\right)$ give the same result.

Also $8 \div 1000$ and 8×10^{-3} $\left(\text{and } 8 \times \dfrac{1}{10^3}\right)$ give the same result.

Write each of these as multiplications, using powers of 10:   a) $5 \div 100$ b) $16 \div 100\,000$.

3 a) Copy and complete each column:

$3.2 \times 10^1 = 32$ $3.2 \times 10^{-1} = 0.32$ $3.2 \div 10^1 = 0.32$
$3.2 \times 10^2 =$ $3.2 \times 10^{-2} =$ $3.2 \div 10^2 =$
$3.2 \times 10^3 =$ $3.2 \times 10^{-3} =$ $3.2 \div 10^3 =$
$3.2 \times 10^4 = 32\,000$ $3.2 \times 10^{-4} = 0.000\,32$ $3.2 \div 10^4 =$
$3.2 \times 10^5 =$ $3.2 \times 10^{-5} =$ $3.2 \div 10^5 =$
$3.2 \times 10^6 =$ $3.2 \times 10^{-6} =$ $3.2 \div 10^6 =$

b) Look at the patterns of your results in a). Explain a quick way of finding the result of:

  (i) multiplying a number by a power of 10 (for example, $1.617\,9 \times 10^7$)
  (ii) dividing a number by a power of 10 (for example, $3\,596.4 \div 10^6$).

4 Write down each result: a) $1.761\,42 \times 10^8$ b) $314\,600 \times 10^{-6}$ c) $516\,494 \div 10^4$.

CHALLENGE

5 Write as single numbers:

  a) $2.45 \div 10^{-3}$ b) $0.0041 \div 10^{-4}$ c) $\dfrac{2 \times 10^{-3}}{2 \div 10^{-4}}$ d) $\dfrac{5 \div 10^{-3}}{5 \div 10^{-8}}$.

TAKE NOTE

This is the result   2.14   For every multiplication by 10
of *multiplying* 2.14   21.4   the digits move one place
successively by 10:   214.0   to the left.
                            2140.0
                            21400.0
                            214000.0

6 Divide 2.14 successively by 10. Write a *Take note* for dividing by 10 like the one for multiplying by 10.

7 Do not use a calculator in this question.

  a) Write each of these as decimals:

    (i) 2.176×10^5 (ii) $2.176 \div 10^5$ (iii) 2.176×10^{-5}
    (iv) $0.002\,15 \times 10^8$ (v) $0.002\,15 \div 10^8$ (vi) $0.002\,15 \times 10^{-8}$

  b) Copy and complete:

    (i) $3.64 \times 10^? = 36\,400$ (ii) $3.64 \times 10^? = 0.003\,64$
    (iii) $3.64 \div 10^? = 0.364$ (iv) $3.64 \div 10^? = 36.4$
    (v) $0.0024 \times 10^? = 24$ (vi) $0.0024 \times 10^? = 0.000\,0024$
    (vii) $0.0024 \div 10^? = 0.000\,24$ (viii) $0.0024 \div 10^? = 0.24$.

B4

━━━━━━ ACTIVITY: EXP ON A CALCULATOR ━━━━━━

7   You need a scientific calculator with an ▮EXP▮ key.

a)   Press ▮C▮ ▮3▮ ▮EXP▮ ▮4▮ ▮=▮

Write down the result.

Try some more examples, for example, ▮C▮ ▮5▮ ▮EXP▮ ▮5▮ ▮=▮

Try some examples using negative numbers such as

▮C▮ ▮5▮ ▮EXP▮ ▮2▮ ▮±▮ ▮=▮ and ▮C▮ ▮2▮ ▮±▮ ▮EXP▮ ▮3▮ ▮=▮

Explain what the EXP key does on a calculator.

8   a) Do not use a calculator in this part.
Write down each result:

(i) 1.7×10^3 (ii) 8.9×10^{-4} (iii) $180\,000 \times 10^{-8}$.

b) Check your results in a) by using the ▮EXP▮ key on your calculator.

━━━━━━ TAKE NOTE ━━━━━━

The key sequence ▮C▮ ▮1▮ ▮.▮ ▮2▮ ▮EXP▮ ▮5▮ ▮=▮ calculates 1.2×10^5

and ▮C▮ ▮1▮ ▮.▮ ▮2▮ ▮EXP▮ ▮5▮ ▮±▮ ▮=▮ calculates 1.2×10^{-5}

9   Do not use a calculator in this question.

A humming bird weighs about 2×10^{-3} kg.
A parasitic wasp weighs about 5×10^{-9} kg.

If a humming bird ate its
own weight in parasitic
wasps, how many wasps
would it eat?

Is it about

A   4000
B   40 000
C   400 000 or
D   4 000 000?

10 Use a ruler to estimate the mean width of the grooves of an LP. Write your estimate using indices, in millimetres (for example, 7×10^{-4} mm).

This scanning electron micrograph of the surface of a long-playing stereo record shows a scratch across the grooves. The grooves are cut into a flat disc of polyvinyl chloride (PVC). The nature of the groove varies with the intensity of the music: the straighter the groove the quieter the music; the more wavy the groove (as here) the louder the music. The magnification is $\times 40$.

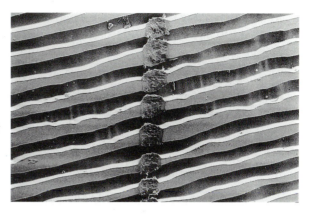

11 a) Use a ruler to measure the lengths shown for the three bacteria. Write your measurements, using indices, in millimetres.

    b) Which of the three bacteria is
        (i) the longest (ii) the widest?

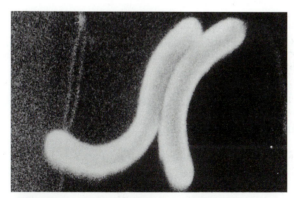

Vibrio cholerae, $\times 10\,000$

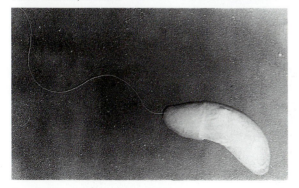

Desulfovibrio desulfuricans, $\times 350\,000$

Responsible for Legionnaire's disease

A *Legionella* bacillus

Legionella pneumophilia, $\times 2000$

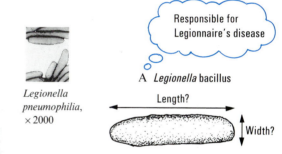

Length?

Width?

B *Cholera* bacterium

Body length?

Width?

C Sulphate-reducing bacterium

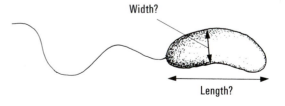

Width?

Length?

B4

ENRICHMENT

1 a) Arrange these in order of size, smallest
       first (do not use a calculator to help you):

$$2^6 \qquad 2^{5.8} \qquad 2^{6.2} \qquad 2^7$$

   b) Check your results in a) with a calculator.

   c) Check that (i) $3^3 = 27$ (ii) $3^4 = 81$.

   d) Which do you think is true?

       $3^{3.5}$ is nearer 27 than 81
       $3^{3.5}$ is nearer 81 than 27
   or $3^{3.5}$ is exactly between 27 and 81.

   e) Sketch the graph. Explain how the shape
       of the graph can help you to answer d).

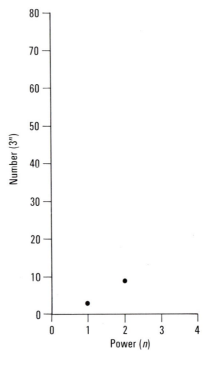

2   In your head estimate the value of x for each equation. Then use your calculator to get a closer
    approximation. Carry on using your calculator until you have each answer to the nearest 0.1.

    a) $3^x = 60$ b) $2.4^x = 100$.

WITH A FRIEND: THE POWERS GAME

3   *Player 1* chooses a base number between 1 and 5, for example, 3.6.

    Player 2 chooses an index between 1 and 5, for example, $3.6^{4.9}$. ⟵ index
                                         ↖ base number

    Each player estimates the result in his/her head and writes it down. Now find the result using
    the ⬛ key and find whose estimate is closer to the true result. The better estimate scores 1 point.
    Take turns to choose another base number and another index. The first player to score 10 points
    wins.

B4

B5 FACTORS AND MULTIPLES

CORE

IN YOUR HEAD

1   Do these in your head. Write down only the answer.

   a) The first multiple of 7 is 7. What is the seventh multiple of 7?

   b) 18 has four factors not counting 1 and 18. What are they?

   c) 22 and 55 share two factors. What are they?

   d) Which number has only one factor?

   e) What is the first number which is both a square number and a triangle number?

   f) Write down the first number which is a multiple of both 4 and 6.

   g) The tenth triangle number is 55. What is the eleventh?

   h) Add together the first four odd numbers. Which square number is this?

EXPLORATION

2   12 has six factors: 1, 2, 3, 4, 6, 12.
    Four of these are even (2, 4, 6 and 12) and two are odd.

   a) Find some numbers which have all their factors even, apart from 1. Describe the sequence of numbers which these form.

   b) Find some numbers which have exactly half their factors even. Describe the sequence of numbers which these form.

   c) Investigate some more combinations of odd and even numbers of factors. Record any interesting sequences you find.

3   a) List the factors of (i) 24 (ii) 42.

   b) List the numbers which are factors of both 24 *and* 42. What is special about the number 6 with respect to 24 and 42?

   c) Find the largest number which divides into both 28 and 42.

4   Copy and complete the *Take note* on the next page.

━━━━━━━━━━━━ TAKE NOTE ━━━━━━━━━━━━

The factors of 30 are 1, 2, 3, ☐, ☐, 10, ☐, 30.
The factors of 45 are 1, ☐, 5, ☐, 15, 45.
The numbers which are factors of both 30 and 45 are 1, ☐, ☐, ☐.
The *highest* common factor (HCF) of 30 and 45 (that is, the largest number which is a factor of both 30 and 45) is 15.

5   a) Write down the factors of (i) 24 (ii) 60.   b) What is the HCF of 24 and 60?

6   Find the HCF of: a) 12 and 16   b) 40 and 100   c) 36 and 136.

7   a) Which factors are common to 12, 28 and 36?

    b) What is the HCF of 12, 28 and 36?

8   a) Write down two numbers which have 7 as their HCF.

    b) Write down three numbers which have 8 as their HCF.

9   If we choose any 10 numbers *less* than 1000, their lowest possible common divisor (factor) is 1. What is their highest possible common divisor?

10 Two blocks are to be built out of 1 cm cubes. One block will have a volume of 36 cm³, the other 60 cm³.

    a) If the blocks are to have the same height, what is the tallest they can be?

    b) Explain what your result has to do with common factors.

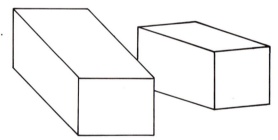

━━━━━━━━━━━━ CHALLENGE ━━━━━━━━━━━━

11 You want to tile a bathroom floor and a kitchen floor with the same style and size of tile. You do not want to cut any tiles; that is, you want to use only whole tiles.

    The kitchen floor measures 3.0 m × 2.7 m and the bathroom floor 2.4 m × 2.8 m.

    a) What size of tiles could you use? (Choose only sensible sizes. Give at least one possibility.)

    b) What is the largest size of tile you could use?

Common multiples

1 A fog-horn sounds every 20 seconds.
   Another fog-horn sounds every 25 seconds.
   They both sound together at midnight.

   a) After how many more seconds do they
      next sound together?

   b) Write down the first five times after
      midnight when they sound together, like
      this:

      00:1:40,   00:3:20, ... (hour:minute:second)

2 These are two fences on opposite
   sides of a road. The first posts
   in each fence are directly opposite
   each other.

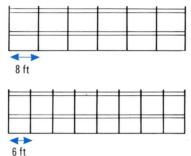

   a) Which are the next two posts which are directly opposite each other?

   b) Write down the distances from the first post, of the next five posts which are opposite each
      other, like this:

      24 ft, 48 ft, ...

3 A number 7 bus leaves the bus station every 18 minutes. A number 2 bus leaves every hour. The
   first two buses of the day leave together at 6:30 am.

   a) When do a number 7 and a number 12 bus leave together again?

   b) How many number 7 and number 12 buses leave the station together between 6:30 am and
      11:30 pm?

4 a) List the first eight multiples of 9.

   b) List the first eight multiples of 12.

   c) List the first three numbers which are multiples of both 9 and 12.

B5

5   These are the multiples of 8 and 12.

8,   16,   (24,)   32,   40,   (48,)   56,   64,   (72,) ...

12,   (24,)   36,   (48,)   60,   (72,)   84,   96,   108, ...

   a) The ringed numbers are multiples of both 8 and 12. Write down the next two examples.

   b) Copy and complete the *Take note*.

▨▨▨▨▨▨▨▨▨▨▨▨▨ TAKE NOTE ▨▨▨▨▨▨▨▨▨▨▨▨▨▨▨▨▨▨▨▨▨▨▨▨▨▨▨▨▨▨▨▨▨▨▨

The first 6 multiples of 12 are 12, 24, ☐, 48, 60, ☐.
The first 6 multiples of 16 are 16, ☐, ☐, 64, 80, ☐.
The first 4 multiples common to 12 and 16 are 48, ☐, ☐, 192.
The lowest (smallest) common multiple (LCM) of 12 and 16 is ☐.

▨▨

6   Write down two numbers which have 16 as their lowest common multiple (LCM).

7   Write down three numbers which have 20 as their LCM.

8   Find the LCM of: a) 2 and 3   b) 5 and 9   c) 6 and 12   d) 9, 15 and 18.

9   In a factory a chair is produced every 8 minutes and a table every 20 minutes.

   a) After how many minutes are the first chair and table finished together?

   b) In an 8 hour day, working continuously, how many tables and chairs will be finished together?

   c) At the end of the day, how many chairs will have been produced for each table?

10 The largest collection of numbers
    less than ten which have 10 as their LCM is   1, 2, 5

    What is the largest collection of numbers less than 100 which share 100 as their LCM?

▨▨▨▨▨▨▨▨▨▨▨▨▨ CHALLENGE ▨▨▨▨▨▨▨▨▨▨▨▨▨▨▨▨▨▨▨▨▨▨▨▨▨▨▨▨▨▨▨▨▨▨▨▨▨▨

11 These are instructions from an old textbook on how to add two fractions.

To add: $\frac{7}{8} + \frac{5}{12}$
Find the LCM of 8 and 12: 24

Use this as a common denominator: $\dfrac{☐ \ + \ ☐}{24}$

24 is 3×8, so multiply 7 by 3.
24 is 2×12, so multiply 5 by 2 $\dfrac{7 \times 3 + 5 \times 2}{24}$ The answer is $\frac{31}{24}$

   a) Check that the result is correct.   b) Explain why the method works.

Prime factors

1   The number 12 can be written as a *product* of *prime numbers* (that is, as a set of prime numbers multiplied together), like this:

$12 = 2 \times 2 \times 3 = 2^2 \times 3$

So can the number 60: $60 = 2 \times 2 \times 3 \times 5 = 2^2 \times 3 \times 5$

a) Write 36 as a product of primes.

b) Write 121 as a product of primes.

c) Write 4848 as a product of primes.

d) Investigate some numbers of your own. Write each one as a product of primes.

e) Which numbers cannot be written as a product of primes?

2   Here is a method for writing 84 as a product of primes.

Represent 84 as a prime multiplied by another number.

Represent the other number as a prime multiplied by a number.

... and so on ...

So 84 is $2^2 \times 3 \times 7$.

Draw a 'tree' diagram to help you to write each of these as a product of primes.

a) 80   b) 120   c) 289   d) 400

3   This is another way of finding the primes which multiply together to make a number.

So 75 is $5^2 \times 3$.

a) Explain how the method works.

b) Use the method to express each of these as a product of primes.

(i) 24   (ii) 64   (iii) 120   (iv) 500   (v) 4840.

4   a) What is the smallest number which has the first ten primes as factors?

b) Write down five more factors of the number.

B5

REVIEW

There are three main ways of indicating what is 'typical' or 'average' about a set of data readings (for example, a set of midday temperatures for Glasgow):

- The most common reading in the set is called the *mode* or *modal value*.
- When the readings are arranged in order (for example, lowest temperature to highest), the middle temperature is called the *median*.
- By addding up all the temperatures and dividing by the number of temperatures in the set we obtain the *mean* ('a fair shares for all' average).

We can refer to each of these, *mode*, *median* or *mean*, as the 'average'.

Here are examples for the set of midday temperatures for Glasgow:

Midday temperatures (°C) in Glasgow during 1–14 June
17 18 17 16 20 21 23 20 17 19 26 27 28 24

Mean: $\dfrac{17+18+17+16+20+21+23+20+17+19+26+27+28+24}{14} = \dfrac{293}{14} \approx 20.9$

So the *mean temperature* is approximately 20.9 °C.

Mode:

Temperature (°C)	16	17	18	19	20	21	22	23	24	25	26	27	28
Frequency	1	3	1	1	2	1	0	1	1	0	1	1	1

The temperature with the highest frequency is 17°C. So the *modal temperature* is 17°C.

Median: 16   17   17   17   18   19   20 | 20   21   23   24   26   27   28

  7 temperatures ←   → 7 temperatures

When the temperatures are arranged in order, the two 20°C readings are in the middle. So the *median temperature* is $\frac{1}{2}(20+20)°C = 20°C$.

Hours of sunshine in Glasgow during 1–7 June
8   5   10   9   9   6   9

With an odd number of readings, there is only one reading in the middle:

5   6   8 $\boxed{9}$ 9   9   10

  ←   →

3 readings | 3 readings   So the *median* number of hours of sunshine is 9.

■ What are the mean and modal numbers of hours of sunshine?

CORE

1 a) Help each other to calculate

- the mean age
- the modal age
- the median age

of the people recorded in the table. (You will be checking your calculations in question 3.)

Ages of people on a European coach tour								
35	30	29	37	41	37	29	38	40
38	40	34	37	58	39	42	58	28
50	22	32	46	37	45	24	29	37
42	35	19	44	44	46	35	45	48
55	34	36	24	32	50	23	35	29

b) Both of you write a short paragraph of not more than thirty words for the local newspaper to give an idea of the ages of the people on the coach, without using the words 'mean', 'median' and 'mode'. Decide between you whose paragraph gives the best idea of the 'age profile'.

2 Find a set of five ages from the table in question 1 whose mean, mode and median are equal.

3 A *frequency table* like this can help organise and reduce the work involved in calculating averages.

a) Check that the first two columns are a correct frequency table for the ages in question 1. Notice that the modal age is 37; it has the highest frequency, namely, 5.

b) Copy the table and complete the third column and the total beneath it.

c) Explain how you can now find the *mean* age directly from your table, and find it.

Frequency table		
Age	Frequency	Frequency × Age
19	1	19
22	1	22
23	1	23
24	2	48
28	1	
29	4	
30	1	
32	2	
34	2	
35	4	
36	1	
37	5	
38	2	
39	1	
40	2	
41	1	
42	2	
44	2	
45	2	
46	2	
48	1	
50	2	
55	1	
58	2	
Totals	45	

B6

━━━━━━━ TAKE NOTE ━━━━━━━

To calculate the median age it helps to add another column called the *cumulative frequency* column.

The cumulative frequency column gives a 'running total' of the frequencies.

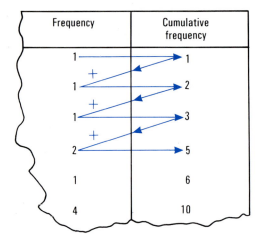

---

d) Add a cumulative frequency column to your table in b). The last entry in your cumulative frequency column should be 45. Why?

e) Explain how you can now read the median age of people on the coach directly from your table, and write down the median age.

━━━━━━━ TAKE NOTE ━━━━━━━

A frequency table (including a cumulative frequency column) helps us to calculate the mean, mode and median of a set of data.

For this set of data,

the *mode* is 4

the *mean* is 2.7 ($107 \div 40$)

the *median* is 3 (the 20th and 21st nests have 3 eggs).

Frequency table: Number of eggs per hen per week

Number of eggs per hen per week	Frequency	Frequency × number of eggs	Cumulative Frequency
0	7	0	7
1	4	4	11
2	3	6	14
3	9	27	23
4	15	60	38
5	2	10	40
Totals	40	107	

4 The *Take note* shows the results of a survey of 40 hens (out of a total of 500) on a farm. You buy 100 hens from the farm. How many eggs would you expect your hens to produce each week? Which 'average' did you use to calculate your estimate, and why?

▨▨▨▨▨▨▨▨▨▨ WITH A FRIEND ▨▨▨▨▨▨▨▨▨▨

5 The table shows the accurate weighing (in grams) of the contents of a sample of 50 yoghurts. Each carton should contain 40 g of yoghurt.

42	42	36	42	41	30	42	43	45	41
41	43	32	43	41	42	33	43	43	42
41	43	31	43	42	41	41	35	37	30
43	43	42	41	42	42	33	41	44	30
31	42	43	41	43	42	43	42	42	35

a) Between you, make out a frequency table, and calculate the mean, mode and median of the weights.

b) Discuss how useful your results in a) are in helping to show that there is a problem with the carton-filling process. Each of you write down what you decide. In your explanation say why one average alone can be very misleading.

▨▨▨▨▨▨▨▨▨▨ THINK IT THROUGH ▨▨▨▨▨▨▨▨▨▨

6 The bar chart shows the length of time (to the nearest five minutes) taken by 40 children on their way to school each day.

a) Explain how you can find the median time from the bar chart, and find it.

b) Explain how you can estimate the mean time from the bar chart, and find it.

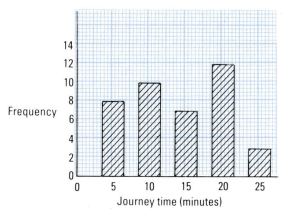

B6

Grouped data

1   A market gardener divided a potato field into eighty equal-sized plots, and used two different types of fertiliser. The numbers show the yields in kilograms of potatoes in each plot; Fertiliser A is shown in blue and Fertiliser B in black.

126	139	143	149	155	126	140	144	149	155
136	140	145	150	155	136	141	146	150	150
137	141	146	151	159	137	141	147	152	159
137	142	147	153	159	138	143	147	154	150
137	140	142	143	143	145	146	148	149	155
136	140	138	142	141	145	146	149	150	150
138	140	141	143	143	146	148	151	152	156
140	141	142	143	145	145	147	148	150	150

To compare the results of the fertilisers he produced two frequency tables.

Fertiliser A

Yield (kg)	Midpoint	Frequency	Midpoint × Frequency	Cumulative Frequency
125–129	127	1	127	1
130–134	132	0	0	1
135–139	137	5	685	6
140–144	142	13	1846	19
145–149	147	11	1617	30
150–154	152	8	1216	38
155–159	157	2	314	40
Totals		40	5805	

Fertiliser B

Yield (kg)	Midpoint
125–129	
130–134	
135–139	
140–144	
145–149	
150–154	
155–159	
Totals	

a) We can *estimate* the mean yield of the plots which used Fertiliser A, like this:

Mean yield $\approx \frac{5805}{40}$ kg $= 145$ kg (3 SF)

Copy and complete the table for Fertiliser B and estimate the mean yield.

b) The median yield for Fertiliser A is about 145 kg. Explain why.

c) Estimate the median yield for Fertiliser B.

d) The modal yield for Fertiliser A is 140–144 kg (or, we might say 142 kg). Explain why.

e) Estimate the modal yield for Fertiliser B.

f) Write a short paragraph comparing the qualities of Fertiliser A and Fertiliser B.

━━━━━━━━━━ TAKE NOTE ━━━━━━━━━━

When there are a large number of different data values, we can save labour by grouping data into intervals. The mean, mode and median can then be estimated from the grouped data tables.

Frequency polygons

The diagram shows the results of a mathematics examination for Fourth Years in a secondary school.

The line joining the tops of the bars is called a *frequency polygon*.
It helps your eye to see how the marks are distributed.

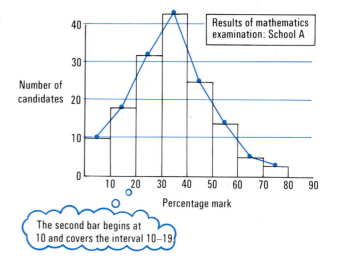

The second bar begins at 10 and covers the interval 10–19.

1 Read the *Take note*.
   a) How many candidates in School A scored:
      (i) between 10 and 19
      (ii) between 40 and 49?
   b) How many candidates sat the examination?
   c) The second chart shows the results in a second school, for the same examination. Redraw one of the charts so that the intervals on both charts are the same. Draw the frequency polygon.
   d) Compare the frequency polygons for schools A and B. Which school has the better examination results, on the evidence of the polygons? Explain how you arrived at your conclusion.

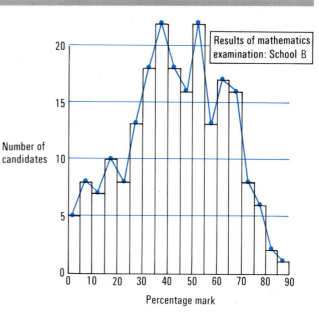

2 For each school, estimate the mean, median and modal score and the range. Do these change your views in question 1 d) about which school has the better results? Explain your answer.

B6

ENRICHMENT

Cumulative frequency graphs

1   The table is a repeat of that
   on page 99 for the ages of
   people on a European coach tour.

   a) You need your frequency
      table from question 3 on
      page 99, and 2 mm graph
      paper. Copy the axes and
      the portion of the graph
      which has been drawn.
      Decide what the points on
      the graph mean, and then
      complete it.

   b) *From your graph* find how
      many members of the
      coach party are less than

      (i)   19 years old
      (ii) 29 years old
      (iii) 50 years old.

      Check from the table that
      you are correct each time.

   c) If you continue the dotted
      arrow to the right until it
      meets the graph and then
      continue it vertically down,
      to meet the 'Age' axis, what
      will the result tell you about
      the 'average' age?

Ages of people on a European coach tour								
35	30	29	37	41	37	29	38	40
38	40	34	37	58	39	42	58	28
50	22	32	46	37	45	24	29	37
42	35	19	44	44	46	35	45	48
55	34	36	24	32	50	23	35	29

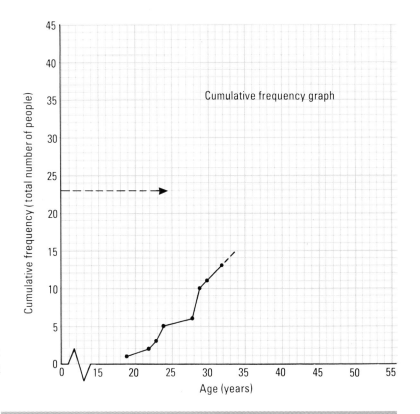

Cumulative frequency graph

TAKE NOTE

The graph you drew in question 1 is called a
cumulative frequency graph. It shows a 'picture'
of the cumulative frequencies when the ages are
arranged in order (along the horizontal axis).
From the graph we can find the median age, by
choosing the 'middle' person on the vertical axis,
and drawing a horizontal and then a vertical line
to the horizontal axis.

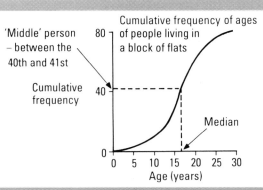

Cumulative frequency of ages of people living in a block of flats

2   You need 2 mm graph paper.
    This is a cumulative frequency graph drawn
    for the yields of potatoes using Fertiliser A
    in question 1 on page 102.

    a) The median yield is the mean of the 20th
       and 21st amounts. Use a set square or
       ruler to estimate the median from the
       graph (continue the dotted line, by eye).
       Check your result with the estimate in
       question 1 b) on page 102.

    b) Draw your own cumulative frequency
       graph for the results of Fertiliser B.
       Estimate the median from your graph,
       and check the result with the result you
       obtained before.

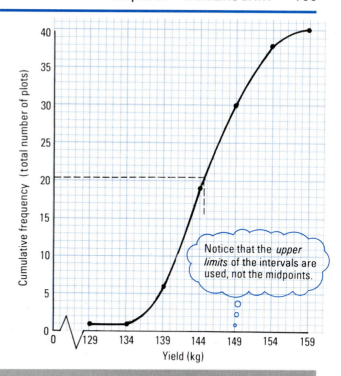

Notice that the *upper limits* of the intervals are used, not the midpoints.

THINK IT THROUGH

3   Why do you think the *upper limits* of the intervals are used for plotting points on a cumulative
    frequency graph for grouped data, and not the midpoints? Write one or two sentences to explain.

TAKE NOTE

Aluminium sheeting				
Hardness (Rockwell's coefficient)	Midpoint	Frequency	Mid. × Freq.	Cumulative Frequency
23–26	$24\frac{1}{2}$	4	98	4
27–30	$28\frac{1}{2}$	10	285	14
31–34	$32\frac{1}{2}$	7	227.5	21
35–38	$36\frac{1}{2}$	7	255.5	28
39–42	$40\frac{1}{2}$	2	81	30
Totals		30	947	

We can *estimate* a *mean* from a table of grouped data, using *midpoints*.

Mean $\approx \frac{947}{30} = 31.6$ (1 DP)

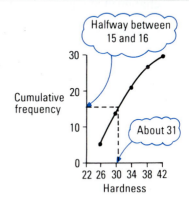

Halfway between 15 and 16

About 31

We can *estimate* a *median* from a table of grouped data by drawing a cumulative frequency graph, using *upper* limits.

Median ≈ 31

4   Here is some information concerning the amounts of milk (in hectolitres, measured to the nearest 0.1 hl) produced by a herd of 220 cows in two different years.

Estimate the mean and median milk yield for each year. Do you think the farmer has reason to be happy or unhappy about the results in 1985 compared with those in 1984?

Milk yield (hl/year)	Number of cows 1984	1985
10.0–14.9	2	0
15.0–19.9	10	10
20.0–24.9	23	25
25.0–29.9	54	60
30.0–34.9	61	58
35.0–39.9	41	44
40.0–44.9	22	20
45.0–49.9	6	3
50.0–54.9	1	0

5   Information is often given using percentages instead of frequencies. The table shows the result of an analysis of skiing accidents at a Swiss resort.

a) Just as we made a cumulative frequency column from a frequency column, we can make a cumulative *percentage* column from a percentage column. Copy the table and add a cumulative percentage column to it. Draw a cumulative percentage graph. (Assume that there were no skiers older than 77.)

Age of skier (years)	Percentage of skiing accidents
under 10	10
10–19	61
20–29	19
30–39	5
40–49	3
50 or over	2

b) Use your cumulative percentage graph to help you *estimate* the median age of skiers involved in skiing accidents at the resort. (When working with percentages, no adjustment is necessary for the fact that the '100' in '100%' is even. Just draw a horizontal line from the 50% mark on the cumulative percentage axis.)

c) In a) you were told to assume that there were no skiers older than 77. Would your result in b) change in any way if you had been told 97 instead? Or 107? Or 57?

d) Comment on the following newspaper report:

The analysis showed that, on the ski slopes, if you make it past your teenage years, then the older you are, the safer you are.

6   The table gives a breakdown of the population of Worthing (in 1975) according to age. Use a cumulative percentage graph to help you estimate the median age of the 1975 population of Worthing.

Age in years	Percentage
under 15	15.7
15–29	14.4
30–44	12.6
45–59	15.7
60–74	25.7
75 or over	15.9

ASSIGNMENT

7   Decide upon two sets of data you would like to collect, from which you can produce a frequency table and calculate or estimate the mean, mode and median to help you to compare the data. Collect the data, and write a report. Illustrate your report with graphs, charts and other drawings which help you to make comparisons.

B7 SOLVING EQUATIONS

REVIEW

To help us to solve problems like this:
we can write a number sentence

$5T+3 = T+19$

and solve it:

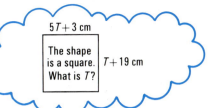

The shape is a square. What is T?
$5T+3$ cm
$T+19$ cm

- by trial and improvement
or
- by using a 'balance' method.

Try $T = 2$ $10 + 3 = 2 + 19$ No
Try $T = 3$ $15 + 3 = 3 + 19$ No
Try T ......

$$5T + 3 = T + 19$$

Subtract 3 from both sides.	-3:	$5T = T + 16$
Subtract T from both sides.	$-T$:	$4T = 16$
Divide both sides by 4.	$\div 4$:	$T = 4$

Check: $\underbrace{5 \times 4 + 3}_{23} = \underbrace{4 + 19}_{23}$

■ Solve this equation: $2+6k = 12-k$
   a) by trial and improvement   b) by using the 'balance' method.

CORE

1   Here are some solutions to equations, using the 'balance' method. Each solution has some parts missing (represented by a '?' each time). Copy the solutions, and fill in what is missing.

a)

	$3k + 1 = 4 - k$
-1:	$3k = ? - k$.
$+ k$:	$?k = 3$
$\div 4$:	$k = ?$

| Check | $3 \times ? + 1 \overset{?}{=} 4 - ?$ |
| | $? = ?$ ✓ |

b)

	$13 + k = 4 - 2k$
$+ 2k$:	$13 + ? = ?$
-13:	$3k = ?$
$?$:	$k = ?$

| Check | $13 + {}^-3 \overset{?}{=} 4 - {}^-?$ |
| | $? = ?$ ✓ |

2   Solve each of these equations using the 'balance' method. Check the solution each time.

   a) $4m-4 = m+6$ b) $2+3n = 5-5n$ c) $t-9 = 4-t$ d) $n-7 = 8-2n$.

Equations with brackets

1 a) Think carefully about this problem. Try
       to solve it without using an equation.

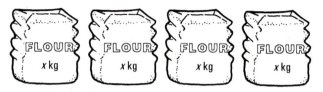

Problem Each bag starts with the same
amount of flour in it. Jim adds 2 kg more
to each bag. Altogether there are now
10 kg of flour. How many kilograms of
flour were there originally in each bag?

   b) Copy and complete this equation for the problem in a):

$$4(x + ?) = 10$$

so $4x + 8 = 10$

-8: $4x = ?$

$\div 4$: $x = ?$

Did you get this answer in part a)?
Check that it fits the problem.

Check $4(? + 2) = 10$
         $4 \times ? = 10$

2 a) Try to solve this problem without using an equation.

Problem Each bag has the same amount
of sugar in it. Rula removes 2 kg from
each. Altogether there are now 12 kg of
sugar. What is p?

   b) Copy and complete this equation for the problem in a):

$$?(p - 2) = ?$$

   c) Copy and complete this solution to the equation in b):

$$?(p - 2) = ?$$

$\div 3$: $p - 2 = ?$

$+ ?$: $p = ?$

Did you get this answer in part a)?
Check that it fits the problem.

Check $3(? - 2) = 12$
         $3 \times ? = 12$

3 Solve each equation. Make sure that you check each of the solutions.

   a) $2(k + 1) = 5$ b) $5(k - 2) = 15$ c) $4(n - 3) = 2$

   d) $2(2a - 1) = 10$ e) $3(1 - a) = 1$ f) $2(4 - 2a) = 2$.

4   This is how Tariq solves the equation $2(n-4) = 10$.

He doesn't use the 'balance' method.

Use Tariq's method to solve each of these equations:

a) $3(p-2) = 9$ b) $2(5-p) = 2$

c) $7(k+1) = 56$.

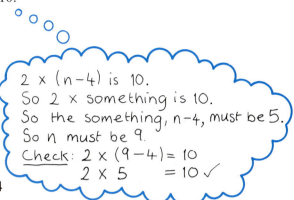

2 × (n−4) is 10.
So 2 × something is 10.
So the something, n−4, must be 5.
So n must be 9.
Check: 2 × (9−4) = 10
          2 × 5   = 10 ✓

5   a) Solve this equation: $6(n-2) = 18$

(i) by trial and improvement
(ii) by using Tariq's method from question 4
(iii) by using the 'balance' method.

b) Solve this equation: $8(n-2) = 3$ using the three methods in a). Write down which method you found easiest and which most difficult.

━━━━━━━━━━━━━━━━ TAKE NOTE ━━━━━━━━━━━━━━━━

There are various methods for solving equations. It is often a good idea to try the easy trial-and-improvement method or Tariq's method first.

━━━

6   Solve each equation. Try trial and improvement or Tariq's method first. If these do not help, use the 'balance' method.

a) $4+2(x-1) = 8$ b) $5+2(x-1) = 9$ c) $3(4-t)-4 = 5$ d) $3(p-2) = p+4$.

━━━━━━━━━━━━━━━━ CHALLENGE ━━━━━━━━━━━━━━━━

7   Copy and complete the solution to each equation:

a) $3(n-4) = 4+2(2-n)$

Multiply out the brackets

$$3n-12 = 4+4-?$$

so $3n-12 = ?-2n$

$+12$: $3n = ?-2n$

?: $5n = ?$

$\div 5$: $n = ?$

Check $3(4-4) = 4+2(2-4)$
$3 \times ? = 4+2 \times {}^{-}?$
$? = 4+?$
$? = 0$

b) $3k+8(1+k) = 20+k$

Multiply out the brackets

$$3k+8+? = 20+k$$

so $?k+8 = 20+k$

-8: $?k = ?+k$

$-k$: $10k = ?$

$\div 10$: $k = ?$

Check $3 \times ?+8(1+?) = ?+?$
$3.6+(8 \times ?) = ?+1.2$
$3.6+? = ?$
$? = ?$

B7

Equations with x^2, x^3, ...

1   The diagram shows a path and a square paved section in a garden. The total area of the path and paved section is 60 m².

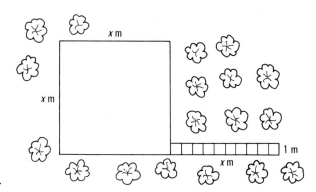

a) Check that this equation satisfies the problem:

$$x^2 + x = 60$$

b) Check that x lies between 7 and 8 (for example, $7^2 + 7 = 56$, so 7 is too small).

c) Use a calculator. Check whether $x = 7.5$ is too small or too large.

d) Make another mental estimate for the value of x. Check with a calculator whether your estimate is too small or too large.

e) Carry on making mental estimates. Find the value of x correct to 1DP.

2   Use a calculator to find x correct to 1DP.

a) $x^2 + x = 30$ b) $x^2 - x = 30$ c) $x^2 = 2x + 40$

d) $x^3 = 57$ e) $x^3 + x = 57$.

WITH A FRIEND: THE EQUATION GAME

3   Choose each equation in turn.
Each of you guess a solution and write it down.

Each of you find the solution correct to 1DP.

Score The one whose guess is closest, gains 1 point. The player with more points out of the possible five, wins the game.

$$x^2 + 2x = 20$$
$$x^2 + 2x = 40$$
$$x^2 + 2x = 60$$
$$x^2 + 2x = 80$$
$$x^2 + 2x = 100$$

More equations

1   a) Try to solve this problem without writing down the equation.

Problem   The total amount of sand in the two piles is packed into t sandbags. Each sandbag then holds 20 kg of sand. How many sandbags are there, and how much sand was there in the second pile?

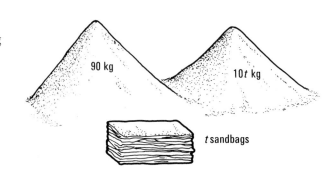

b) Copy and complete this equation for the problem in a):

$$\frac{90+?}{t} = ?$$

c) Copy and complete this solution to the equation in b):

$$\frac{90+?}{t} = 20 \qquad\qquad Check \quad \frac{90+10\times?}{?} = 20$$

$\times t$: $90+? = 20t$

$\qquad\qquad\qquad\qquad\qquad \frac{90+?}{?} = 20$

$-10t$: $90\ = ?$

$\div 10$: $?\ = t$ $\qquad\qquad\qquad\qquad ? = 20$

Does this agree with your answer in a)?

2 a) Try to solve this problem without writing down an equation.

Problem Ellis uses 35 kg of sand from the pile to make mortar. He packs the remainder into t sandbags. Each sandbag now holds 8 kg. How many sandbags are there, and how many kilograms of sand were there originally?

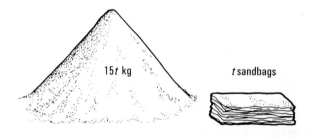

15t kg t sandbags

b) Copy and complete this equation for the problem in a):

$$\frac{?-35}{t} = ?$$

c) Copy and complete this solution to the equation in b):

$$\frac{?-35}{t} = ? \qquad\qquad Check \quad \frac{15\times?-35}{?} = 8$$

$\times t$: $?-35 = 8t$

$\qquad\qquad\qquad\qquad\qquad \frac{?-35}{?} = 8$

$-8t$: $?t-35 = ?$

$+35$: $?t = ?$ $\qquad\qquad\qquad\qquad \frac{?}{?} = 8$

$?$: $t = ?$

Does this agree with your answer in a)?

3 Solve each equation:

a) $\dfrac{6+k}{k} = 4$ $\qquad$ b) $\dfrac{n-3}{n} = 5$ $\qquad$ c) $\dfrac{2d+1}{d} = 4.$

B7

4 a) Try to solve this problem without writing an equation.

Problem Elsie packed m kg of sand into 7 bags, and $(m+5)$ kg of sand into 12 bags. All the bags contained the same amount of sand. How much sand was there in the smaller pile?

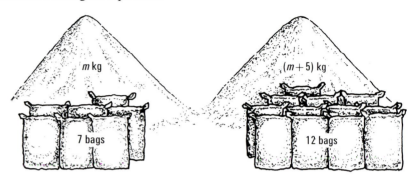

m kg $(m+5)$ kg

7 bags   12 bags

b) Copy and complete this equation for the problem in a):

$$\frac{m}{7} = \frac{?}{12}$$

c) Copy and complete this solution to the equation in b):

$$\frac{m}{7} = \frac{m+5}{12}$$

$\times 7$: $$m = \frac{?(m+5)}{12}$$

$\times 12$: $12m = ?(m+5)$

Remove the brackets:

$12m = 7m + ?$ *Check* $\dfrac{?}{?} = \dfrac{?+5}{12}$

$-7m$: $5m = ?$ $\dfrac{?}{7} = \dfrac{?}{12}$

$\div 5$: $m = ?$

Did you get this answer in a)? Check that the answer fits the problem.

5 a) Try to solve this problem without writing an equation.

Problem Rula cuts the 24 cm and the 40 cm rod into pieces all the same length. She gets 4 more pieces out of the 40 cm rod than she gets out of the 24 cm rod. Into how many pieces does she cut each rod?

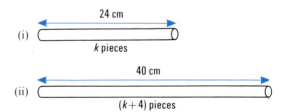

(i)   24 cm
       k pieces

(ii)   40 cm
       $(k+4)$ pieces

b) Copy and complete this equation for the problem in a):

$$\frac{24}{?} = \frac{?}{k+4}$$

c) Copy and complete this solution to the equation in b):

$$\frac{24}{k} = \frac{?}{k+4}$$

$\times k$: $\qquad 24 = \dfrac{40?}{k+4}$

$\times (k+4)$: $24(k+4) = 40k$

Remove the brackets:

$$24k + ? = 40k \qquad\qquad Check \quad \frac{24}{?} = \frac{40}{?+4}$$

$-24k$: $\qquad ? = ?k$ $\qquad\qquad\qquad \dfrac{24}{?} = \dfrac{40}{?}$

$\div 16$: $\qquad ? = k$

Did you get this answer in a)? Check that the answer fits the problem.

6 Solve each of the equations (i) by trial and improvement (ii) by using the 'balance' method.

a) $\dfrac{t}{2} = \dfrac{t+2}{3}$ b) $\dfrac{2}{k} = \dfrac{3}{k+1}$ c) $\dfrac{p}{5} = \dfrac{p+4}{7}$

d) $\dfrac{12}{n} = \dfrac{8}{n-1}$ e) $\dfrac{k+1}{2} = k-5$ f) $\dfrac{5}{m-1} = \dfrac{1}{2}$

7 Solve each of these equations. First try trial and improvement or Tariq's method (page 109). If you are not successful, try the 'balance' method.

a) $3t = 12$ $\qquad$ b) $2t+3 = 9$ $\qquad$ c) $\dfrac{t}{4} = 5$

d) $5-k = 3\frac{1}{2}$ $\qquad$ e) $2(t-1) = 8$ $\qquad$ f) $\dfrac{36}{t} = 9$

g) $3(1+m) = 18$ $\qquad$ h) $5 = \dfrac{t}{6}$ $\qquad$ i) $11-(t-2) = 9$

j) $3m+1 = m+6$ $\qquad$ k) $6k-1 = k+3$ $\qquad$ l) $\dfrac{1}{p}+2 = 2.25$

m) $\dfrac{p}{1+p} = \dfrac{3}{4}$ $\qquad$ n) $\dfrac{1}{t} = \dfrac{4}{t+1}$ $\qquad$ o) $2(1-t) = 1+3t$

p) $\dfrac{4}{m} = \dfrac{6}{m+1}$ $\qquad$ q) $2(n-3) = 4-n$

8 Try to solve $\dfrac{1}{k} = \dfrac{1}{k+1}$

What happens? What does this tell you about the equation?

ENRICHMENT

Inequalities

1   t can represent any number.
     Only one value of t fits this sentence:

   a) $t+4=9$

      Which one?

   b) Many (an infinite number of) values of t fit this sentence:

      $t+4>9$ ($t+4$ is greater than 9)

      Which values?

   c)   (i) Check that the values of t which fit this sentence are less than or equal to 3 ($t \leqslant 3$).

           $9+t \leqslant 12$ ($9+t$ is less than or equal to 12)

       (ii) What is the largest value of t which satisfies the sentence?
      (iii) What is the least value of t which satisfies the sentence?

2   a) Is this true or false:

      If $x>10$ then $^-x>{}^-10$?

      Give examples to explain what you decide. If you decide 'false'
      write down a correct statement involving ^-x.

   b) Repeat a) for the statement:

      If $x>{}^-4$ then $^-x>4$

3   The picture shows an air machine which blows light table-tennis
     balls up and down a closed 50 cm tube, standing on a 40 cm base.

     Explain what each of these inequalities tell us:

   a) $x+40 \leqslant 90$ b) $x \leqslant 50$.

4   Find the solutions to
     these equations and
     inequalities (you should
     be able to solve each
     one by thinking and
     checking):

   a) $x-4=9$ $x-4<9$ $x-4>9$

   b) $2x=6$ $2x>6$ $2x<6$

   c) $\dfrac{x}{2}=3$ $\dfrac{x}{2}>3$ $\dfrac{x}{2}<3$

   d) $\dfrac{40}{x}=5$ $\dfrac{40}{x}>5$ $\dfrac{40}{x}<5$

   e) $\dfrac{1}{x}+4=6$ $\dfrac{1}{x}+4>6$ $\dfrac{1}{x}+4<6$

   f) $2(3+x)=8$ $2(3+x)>8$ $2(3+x)<8$.

WITH A FRIEND

5   These *inequalities* and number-line diagrams are connected:

$x \geqslant {}^-3$

$x < {}^-3 \text{ or } x > 3$

${}^-3 < x < 3$

$x < {}^-3 \text{ or } x \geqslant 3$

${}^-3 \leqslant x < 3$

$x > 2$

$x \leqslant 3$

$x > {}^-3$

$x < 3$

$x \leqslant {}^-3 \text{ or } x \geqslant 3$

${}^-3 \leqslant x \leqslant 3.$

a) Decide between you which inequality fits each diagram and why.

b) Continue working together. Represent the range of values which x can take to satisfy each of these inequalities, on a number-line diagram as in A to K:

   (i) ${}^-3 < 3x \leqslant 3$
   (ii) ${}^-4x > 12$
   (iii) ${}^-6 \leqslant 2x < {}^-1$
   (iv) ${}^-5 \leqslant x + 2 < 7.$

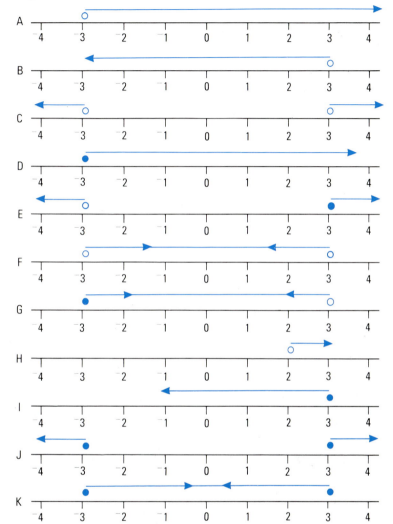

6   The diagram shows the beginnings of a garden path made from paving stones 0.3 m long. The path can be between 12.4 m and 14.9 m long. Use n for the number of paving stones.

a) Decide between you how to write an inequality which tells us how many paving stones can be used.

b) Write down the largest and the smallest values which n can take.

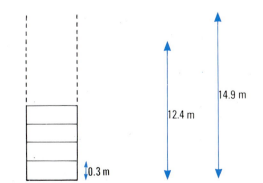

7 a) Explain the connection between the inequality and the number-line diagram.

$$2 < \frac{12}{k} < 4$$

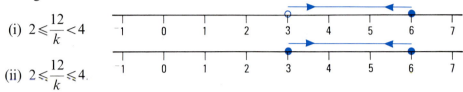

   b) The diagram and the inequality are changed slightly. Explain the connection between the changes.

      (i) $2 \leqslant \frac{12}{k} < 4$

      (ii) $2 \leqslant \frac{12}{k} \leqslant 4.$

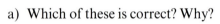

   c) Draw a diagram to represent the values which k can take in the inequality

$$1 < \frac{12}{k} \leqslant 6$$

8 The diameter (D cm) of the circular table is 83.4 cm (to 1 DP).

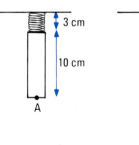

   a) Which of these is correct? Why?

      $83.35 < D \leqslant 83.45$
      $83.35 < D < 83.45$
      $83.35 \leqslant D \leqslant 83.45$
      $83.35 \leqslant D < 83.45.$

   b) Represent the range of possible values for D on a number-line diagram.

9 The diagram shows a 10 cm long metal weight attached to a spring. When the spring is fully compressed it is 3 cm long. When the weight is released from the fully compressed position it extends the spring to 20 cm before it begins its return journey. A is the bottom of the weight.

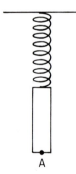
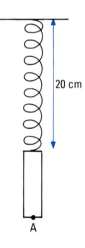

   a) Explain what x represents in this inequality:
      $$13 \leqslant 10 + x \leqslant 30$$

   b) What information does the inequality give us about the point A?

   c) Copy and complete this inequality, and explain in words what it tells us about the weight and the spring:

      $$\square \leqslant x \leqslant \square$$

B7

Solving an inequality

1 Copy and complete each of these inequality solutions:

a) $13 \leqslant x + 10 \leqslant 30$

     $-?$: $? \leqslant x \leqslant ?$

b) $0.6 \leqslant p - 7 \leqslant 10$

     $+?$: $? \leqslant p \leqslant ?$

c) $2p + 1 < 9$

     -1: $2p < ?$

     $\div 2$: $p < ?$

d) $2k - 3 \geqslant 11$

     $+3$: $2k \geqslant ?$

     $\div 2$: $k \geqslant ?$

2 What is the least value which k can take to satisfy each inequality:

a) $8k - 13 \geqslant 13$ b) $18 - 3k \leqslant 3$?

3 Solve each inequality: a) $5n + 7 \geqslant 32$ b) $4 - 2n \leqslant 12$.

4 a) Copy and complete the solution to the inequality:

$$^-16 \leqslant 3n + 1 \leqslant 16$$

     $^-16 \leqslant 3n + 1$ $3n + 1 \leqslant 16$

     -1: $^-? \leqslant 3n$ -1: $3n \leqslant ?$

     $\div 3$: $? \leqslant n$ $\div 3$: $n \leqslant ?$

$$? \leqslant n \leqslant ?$$

b) One solution is correct and one solution is incorrect. Which is which, and why is one wrong?

A $4 - 2x \geqslant 9$ B $4 - 2x \geqslant 9$

    -4: $^-2x \geqslant 5$ -4: $^-2x \geqslant 5$

    $\times ^-1$: $2x \geqslant ^-5$ $\times ^-1$: $2x \leqslant ^-5$

    $\div 2$: $x \geqslant \dfrac{^-5}{2}$ $\div 2$: $x \leqslant \dfrac{^-5}{2}$

c) Solve each inequality and represent the solution on a number-line diagram:

(i) $7n < 12 - 5n < n$ (ii) $^-12 < \dfrac{3}{n} < ^-9$ (iii) $n - 3 < 2n + 5 \leqslant 3n + 4$ (iv) $\dfrac{1}{n} < \dfrac{2}{3n} < \dfrac{6}{7n}$

════════ **TAKE NOTE** ════════

The *solution* of the *inequality* $4 < p + 4 \leqslant 9$ is $0 \leqslant p \leqslant 5$.
We can use the same methods for solving inequalities that we used in equations:

- trial and improvement
- the 'balance' method.

B7

CORE

1 This is the front view of a semi-circular aircraft hangar. Approximately how far is it around the curved part of the hangar? (Use $\pi \approx 3.14$.)

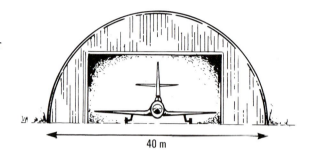

40 m

2 This is a plan of a circular art gallery.

There is a video camera at the centre which sweeps through a 60° arc.

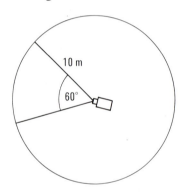

10 m

60°

a) What fraction of the distance around the gallery does the camera survey? Is it $\frac{1}{60}$, $\frac{6}{10}$ or $\frac{1}{6}$?

b) What is the circumference of the art gallery? (Use $\pi \approx 3.14$.)

c) How many metres of the wall does the camera sweep through?

3 This is the equipment for a swing-ball game. As the players hit the ball it swings around the pole, in circles. The string is 2 m long.

a) Roughly, what is the greatest distance the ball travels in one complete circuit? (Use $\pi \approx \frac{22}{7}$.)

b) The ball takes 2 seconds to complete the circuit. What is its average speed?

4   This flexible sheeting is used for the sides of a circular garden pool. Roughly, what is the diameter of the pool? (Use π ≈ 3.1.)

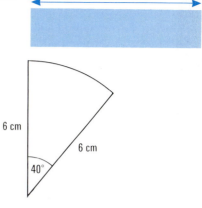

15 m

5   This is a 40° sector of a circle of radius 6 cm.

   a) What is the circumference of the full circle? (Use π ≈ 3.14.)

   b) What fraction of the area of the circle is the sector?

   c) What is the distance along the curved edge of the sector?

   d) Another sector of the circle has its curved edge 12 cm long. Roughly, what fraction is this of the circumference?

   e) Roughly, what is the angle of the sector in part d)?

6 cm

6 cm

40°

CHALLENGES

6

This spin dryer turns at the rate of 2000 revs (revolutions) per minute for 3 minutes. The radius of the drum is 0.3 m. Roughly, how far do your socks travel during each 3 minute spin?

Welded edge

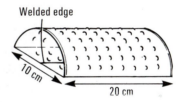

10 cm

20 cm

7   A cheese grater is to be made in one piece from perforated metal sheet like this:
   Roughly what length of metal sheet is needed?

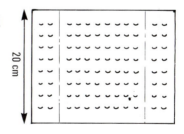

20 cm

8   Make a cone whose base circumference is 20 cm.

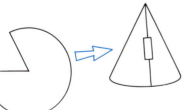

9   The lefthand pulley wheel is driven by a motor. The belt drives the righthand wheel. Each wheel has a radius of 0.2 m, and their centres are 1.4 m apart. Roughly, how long is the belt?

B8

10 Design a drum which will hold 1000 metres
of steel rope with a diameter of 6 cm.

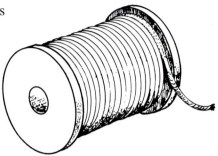

Area of a circle

1   The area of the circle is a little more than the area of the inside
square and a little less than the area of the outside square.

a) Find the area of each square.

Hint
The area of the outside   and the area of the
square is four times the   inside square is four times
shaded area: this shaded area:

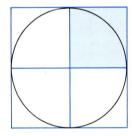   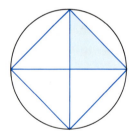

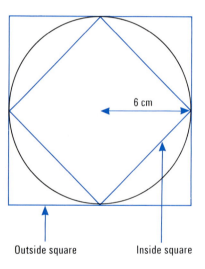

Outside square   Inside square

b) Copy and complete:

$\square$ cm^2 < Area of the circle < $\square$ cm^2

2   Copy and complete the sentence for this circle:
Use the same method as in question 1.

$\square$ cm^2 < Area of the circle (cm^2) < $\square$ cm^2

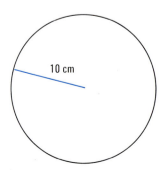

10 cm

3   Copy and complete the sentence for this circle:
    Use the same method as in questions 1 and 2.

$\Box \times r^2$ cm^2 < Area of the circle (cm^2) < $\Box \times r^2$ cm^2

The *approximate* area of a circle, radius r cm, is $3r^2$ cm^2 ($3 \times r^2$ cm^2).
The *exact* area of a circle, radius r cm, is πr^2 cm^2 ($\pi \times r^2$ cm^2).

4   Use the rule: area of a circle $\approx 3 \times r^2$.

a) Estimate the area of turf needed for the circular lawn, radius 3 m.

b) Estimate the area of lawn the sprinkler waters (distance across the circular watered area = 5 m).

5   In this design the largest circle has a radius of 4 cm. Compare the areas of the three circles. How many times larger than the area of the smallest circle is the area of

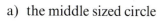
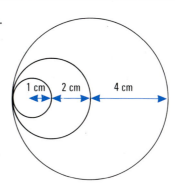

a) the middle sized circle

b) the largest circle?

B8

6   To work out the cost of materials for the
    table mat Jim needs to know its surface area
    as accurately as possible.

a) Which approximation for π should he
   use:

   $\pi \approx 3$, $\pi \approx 3.1$,   or $\pi \approx 3.14$?

b) Calculate the area yourself, using the
   three approximations for π. What is the
   largest difference the three values give?

c) What is the area of the table mat to the nearest 1 cm²?

7   What is the area of this
    sector of a circle?
    (Use $\pi \approx 3.1$.)

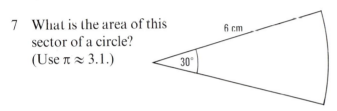

ENRICHMENT

1   The area of the base of the cylinder is πr^2 cm².
    A cylinder is a special kind of prism. Its volume is therefore
    'base area × height', or $\pi r^2 h$ cm³.

a) Find the volume correct to 1 DP if r is 4 cm and h is 5 cm.
   (Use $\pi = 3.142$.)

b) h doubles, but r remains the same. What happens to the
   volume?

c) r doubles but h remains the same. What happens to the
   volume?

d) The volume of a cylinder 10 cm tall is 150 cm³.

   (i) Approximately what is the radius?
   (ii) Calculate the radius correct to 1 DP.
   (Use $\pi = 3.142$.)

e) The diagram shows a net for the cylinder.
   Use it to explain why the surface area of
   the cylinder is $\pi r^2 + \pi r^2 + 2\pi r h$ cm².

f) Calculate the surface area if r is 4 cm and
   h is 5 cm. (Use $\pi = 3.142$.)

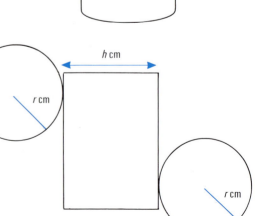

B8

2   p cm is the length of a rectangle. q cm is its width.

   a) (i) What does pq represent? (the perimeter? the area? ...)
      (ii) What units should be used for pq? (cm, cm^2, cm^3, ...)

   b) (i) What does $p+q$ represent?
      (ii) What units should be used for $p+q$?

3   The cuboid has height k cm, length t cm and width n cm.
     Three of the faces have been named A, B and C.

     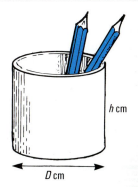

     Explain what each of these expressions represent, and say
     what units should be used for each one (cm, cm^2, cm^3, ...):

   a) kn b) $2(k+n)$ c) $t+k+n$ d) $2tn$

   e) tkn f) $\frac{1}{2}tkn$ g) $2(tk+kn+tn)$ h) $4(t+k+n)$.

━━━━━━━━━━━━━━━━ CHALLENGE ━━━━━━━━━━━━━━━━

4   The pencil holder has diameter D cm and height h cm.
     Decide what each of these expressions represent (for example,
     capacity of the pencil holder, area of curved surface, area of
     base, etc.). For each expression say what units should be used.

   a) πDh b) $\frac{1}{4}\pi D^2$ c) $\frac{1}{4}\pi D^2 h$ d) $\pi Dh+\frac{1}{4}\pi D^2$.

B8

B9 DRAWING SHAPES

B9 CORE

ACTIVITY: DRAWING PERPENDICULAR LINES

1   You need a pair of compasses and a ruler.

a) Draw any circle. Now draw any
   other smaller circle to cut (intersect) the first circle.

b) Join the centres of the
   circles, and the points
   where they cross.

c) Check that the two
   lines are
   perpendicular.

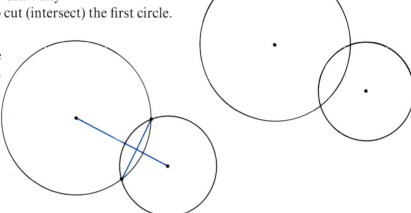

THINK IT THROUGH

2   You need a pair of compasses and a ruler.

a) Here is an explanation of how to draw a line perpendicular to another. Follow it.

(i) Draw a line.

A ———————— B

(ii) With your compass
     point on A,
     mark arcs above
     and below AB.

A ———————— B

(iii) With your compass
      point on B, mark
      two more
      arcs to cut   A ———————— B
      the first arcs.
      (You can change the
      stretch of your compasses or
      leave them as they are)

(iv) Join the two points
     where the arcs
     intersect.

A ———————— B

b) Can you explain how the method works?

======= ACTIVITY =======

3 You need a pair of compasses and a ruler.

a) Draw any circle. Now draw another circle, *the same size*, which cuts the first circle.

b) Join the centres of your circles, and join the points where the circles intersect.

c) Check that the two lines are perpendicular.

d) Explain what is special about the two lines which is not special when the circles are a different size.

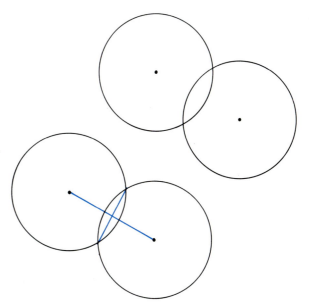

======= THINK IT THROUGH =======

4 You need a pair of compasses and a ruler.

a) Here is an explanation of how to draw the *perpendicular bisector* of a line.

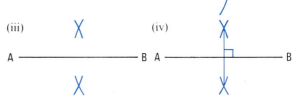

A line which cuts another line in half and at right angles.

Perpendicular bisector of AB

(i) Draw a line AB.

(ii) With your compass point at A, mark arcs above and below AB.

(iii) Keep your compasses open the same amount. With your compass point at B, mark two more arcs to intersect the others.

(i)

A ——————————— B

(ii)

A ——————————— B

(iii)

A ——————————— B

(iv)

A ——————————— B

(iv) Join the two points where the arcs intersect. This is the perpendicular bisector of AB.

b) Can you explain how the method works?

B9

5   You need some plain paper, a pair of compasses and a ruler.
     Draw a square with sides 6 cm long.

ACTIVITY: DRAWING A TRIANGLE

6   You need a pair of compasses and a ruler.

   a) Draw a line 7 cm long.
      Call it AB.

   b) With centre A draw a
      circle of radius 5 cm.
      With centre B draw a
      circle of radius 4 cm.
      Mark C as shown.

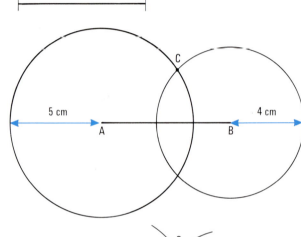

   c) Join AC and BC.

   d) ABC is a triangle. How
      long is (i) AC (ii) BC?

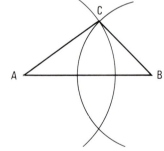

   e) Use the method in a) to d) to draw a triangle whose sides are 3 cm, 4 cm and 5 cm long.

   f) Use the method to try to draw a triangle whose sides are

      (i) 5 cm, 2 cm and 3 cm long
      (ii) 5 cm, 2 cm and 2 cm long.

      What happens?

THINK IT THROUGH

7   One side of a triangle is 8 cm long. What can you say about the lengths of the other two sides?

8   This is an explanation of how to draw a triangle whose sides are 6 cm, 4 cm and 3 cm long:

a) Draw a 6 cm line AB.

b) With centre A draw an arc of radius 4 cm.

c) With centre B draw an arc of radius 3 cm. Join up ABC.

Can you explain how the method works?

▓▓▓▓▓▓▓▓▓▓▓▓▓ CHALLENGE ▓▓▓▓▓▓▓▓▓▓▓

9   a) You can use any of these: a pair of compasses, a protractor, a ruler. Draw these triangles:

(i)

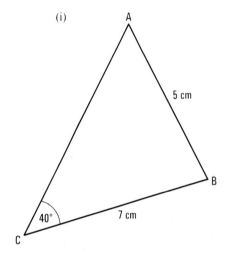

(ii)

(iii)

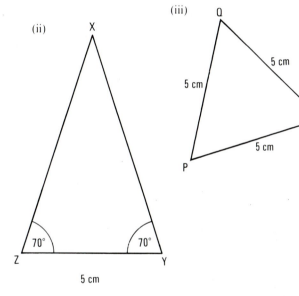

(iv)

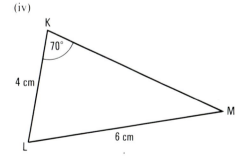

(v)

b) Measure the lengths of sides and the angles of each triangle. Write them on your drawings.

ACTIVITY: BISECTING ANGLES

10 a) Draw any circle, centre
     A, and mark any two
     points B and C on its
     circumference.

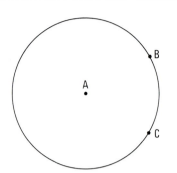

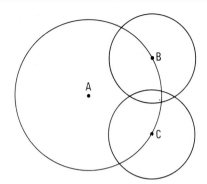

   b) Draw two equal
     intersecting circles with
     B and C as centres.

   c) Your completed diagram has a line of symmetry. Draw it.

   d) Join AB and AC.

   e) What can you say about angles BAX and CAX?

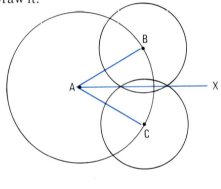

   f) Draw any angle A.

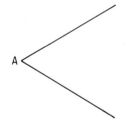

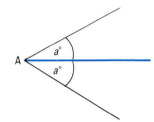

        By drawing three circles, bisect your angle into two equal angles.

11 This is an explanation of how to draw an angle bisector. Follow the instructions yourself.

   a) Draw any angle A.

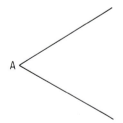

   b) With centre A, draw
     two arcs at B and C.

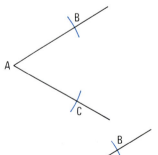

   c) With centres B and C
     and the same radius,
     draw intersecting arcs
     to meet at X like this.

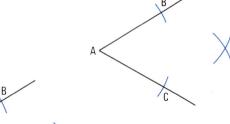

   d) Join AX. This is the angle bisector.

   Can you explain why the method works?

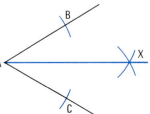

ENRICHMENT

▨▨▨▨▨▨▨▨▨▨ ASSIGNMENT ▨▨▨▨▨▨▨▨▨▨

1 a) Find methods of drawing the different shapes on this page. For each shape write your own instructions to help someone else to draw the shape (see question 8 on page 127).

You may use ● a pair of compasses
            ● a ruler
            ● a set square

(i) This rhombus with 4 cm sides and the shorter diagonal 3 cm long.

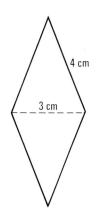

(ii) This trapezium with two right angles.

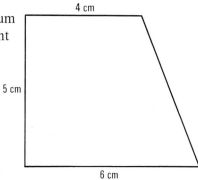

(iii) This parallelogram, with the shorter diagonal 5 cm long.

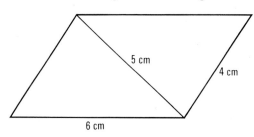

(iv) This kite.

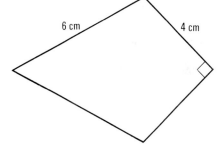

b) Choose a shape of your own which you would like to draw. Draw it accurately. Write instructions to help someone else to draw it.

B10 RATIO AND PROPORTION

REVIEW

- The *ratio* of small buttons to large buttons on the card is 3:1 (6:2, 12:4, 24:8, etc.).

■ Write the ratio 20:16 in three more ways.

- The mass of metal needed to make a metal rod increases *in proportion* to the length of the rod. For example, three times the length weighs three times as much, and so on.

CONSOLIDATION

1 Which of these vary in proportion to each other?

a) A person's height and weight as the person gets older.

b) The amount of water in a bath and the time for which the taps are on.

c) The number of eggs in a saucepan and the time they take to boil.

d) The time I stand on an escalator and the vertical distance I travel.

e) The diameter of a tomato and its weight.

2 Write these ratios in the form ☐:1.

a) 24:2   b) 9:5   c) 3:8.

3 The scale of a map is 20 000:1. How many kilometres does each centimetre on the map represent?

4 171 mothers with just one child (under 10 years old) were asked 'Do you go out to work?' The table shows the results.

	Working mothers	Non-working mothers
0–4 year-olds	15	38
5–9 year-olds	28	90

a) Write each of the ratios below in the form ☐:1.

(i) working mothers of 0–4-year-olds : total number of mothers of 0–4-year-olds

(ii) working mothers of 5–9-year-olds : total number of mothers of 5–9-year-olds

b) Which group of mothers tends to go out to work more, mothers of 0–4-year-olds or mothers of 5–9-year-olds? Explain how you decided.

Ratio division

B10

1   You need a coin and 10 beans (or paper clips or matchsticks, etc.) each, to represent points. The aim of the game is to capture all of your opponent's points. You win when your opponent has no points.

Rules

Decide who will go first. (You can take turns to start.)

● Player 1 chooses a ratio (say 3 : 2) into which Player 2's points are to be divided, and divides Player 2's points into two piles according to the ratio. Note: neither number in the ratio can be larger than 10.

● Player 1 spins the coin. *Heads* means *win*. If Player 1 wins he/she captures the larger of the two points piles from Player 2. If Player 2 wins, Player 1 forfeits that number of points to Player 2.

● Now it is Player 2's turn. Player 2 nominates a ratio for dividing Player 1's points and the game continues as previously.

Note   If a Player chooses a ratio into which the opponent's points pile cannot be divided, the turn is forfeited. (You must therefore keep a record of the number of points your opponent has.)

For example,

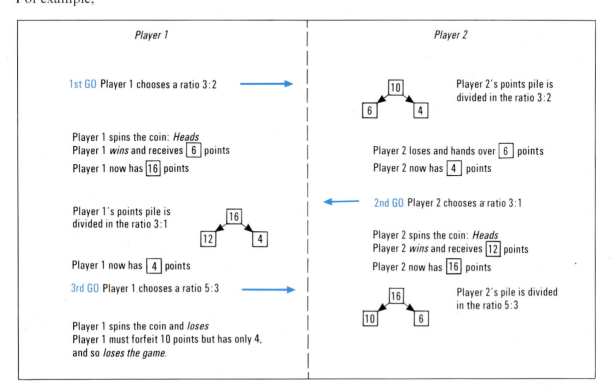

TAKE NOTE

divided in the
ratio 3 : 5 gives

16 cm

$\frac{3}{8}$ of
16 cm

6 cm

10 cm

$\frac{5}{8}$ of
16 cm

£300 shared in the ratio 7 : 3

gives

£210 and £90

$\frac{7}{10}$ of £300

$\frac{3}{10}$ of £300

1 The bottling machine sends two bottles down track A to be filled with orange drink for every three bottles it sends down track B to be filled with lemon drink. So it selects bottles in the ratio 2:3.

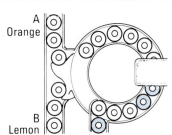

a) In every 100 bottles, how many are filled with orange and how many with lemon?

b) During a day 3680 bottles are filled. How many are orange and how many are lemon?

c) One week 18 600 bottles of lemon are produced. How many bottles of orange are produced?

2 a) Out of every 400 people at an airport, customs officers search about 16. Write this (i) as a ratio of people searched to people not searched (ii) as a percentage.

b) On a certain day 180 people are searched. Roughly how many are not searched?

c) During one week 76 800 people pass through customs. Roughly how many are searched?

d) Due to staff shortages one week, it is decided that only three per cent will be searched. Write the ratio of people searched to people not searched in the form 1 : ☐. How many are not searched if 1400 *are* searched?

3 a) The ratio of the areas of two lawns is about 3:5. A 3 kg bag of lawn feed is just enough for both of them. How many kilograms are needed for each?

b) These are scale drawings of two other lawns. A 5 kg bag of lawn feed is enough for both. How many kilograms would you use for each?

CHALLENGE

4 Sand is 20p per kilogram. Cement is 35p per kilogram.
I can buy 10 kg bags of ready-mixed 'sand and cement' for £3. The mixture is 5 parts sand and 3 parts cement. Which is cheaper, mixing my own or buying ready-mixed, and by how much?

Direct and inverse proportion

1   The graphs represent how two quantities vary together. For example, in A we can see that as b increases in value, a also increases in value. £b might represent the cost of hiring a taxi and a minutes the time spent in it (notice that there is an initial hire charge – when a is 0 b is not 0). Discuss between you how the quantities a and b vary together in the situations B–F. For each graph, describe a situation that could have given rise to it, and explain what a and b represent in your example.

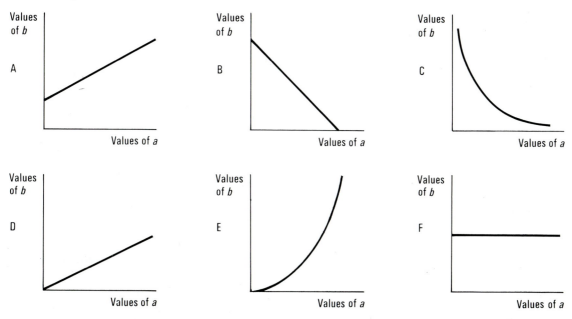

B10

2   Match one of these situations with each graph in question 1.

a)   Quantity a :   the number of millilitres of toothpaste used.
Quantity b :   the number of millilitres of toothpaste left in a tube.

b)   Quantity a :   the number of seconds the conker is swinging in a circle.
Quantity b :   the number of centimetres a conker is from your hand as you swing it in a circle.

c)   Quantity a :   the number of hours you take to travel between two places.
Quantity b :   the number of kilometres per hour at which you travel from one place to another.

d)   Quantity a :   the total number of pounds you pay for 3 m² of curtain material.
Quantity b :   the number of pounds it costs to buy 1 m² of the material.

e)   Quantity a :   the speed at which a ball is thrown into the air.
Quantity b :   the vertical distance it reaches above the ground.

f)   Quantity a :   the number of days hire of a minibus.
Quantity b :   the number of pounds it costs to hire the minibus per day.

3   a) Find the graphs in question 1 for which these are true:

   (i) the values of *a* and *b* increase uniformly (that is, for equal increases in the value of *a* the increases in the value of *b* are equal) (two graphs)
   (ii) as *a* increases in value *b* decreases in value (two graphs)
   (iii) as *a* increases in value, the value of *b* is constant (one graph).

   b) Sketch a graph for which the values of *b* and *a* increase together, but not uniformly.

   c) Find the graphs in question 1 for which these are true:

   (i) $\dfrac{\text{value of } a}{\text{value of } b}$ = constant (one graph only)
   (ii) value of *a* × value of *b* = constant (one graph only)
   (iii) value of *a* | value of *b* = constant (one graph only)
   (iv) value of *b* = constant (one graph only).

━━━━━━━━━━ WITH A FRIEND ━━━━━━━━━━

4   Think of your own examples of two quantities *a* and *b* for which:

   a) the values of *a* and *b* increase together uniformly, *and* $\dfrac{\text{value of } a}{\text{value of } b}$ = constant

   b) *a* increases in value whilst *b* decreases in value, *and* value of *a* × value of *b* = constant.

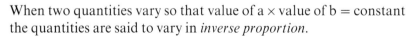

━━━━━━━━━━ TAKE NOTE ━━━━━━━━━━

When two quantities vary so that $\dfrac{\text{value of } a}{\text{value of } b}$ = constant

the quantities are said to vary in *direct proportion*. The graph representing the variation of the two quantities is a straight line passing through (0,0).

For example, the *amount* of cheese you buy and the total *cost* vary in direct proportion.

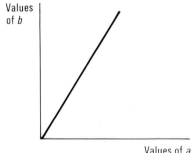

When two quantities vary so that value of *a* × value of *b* = constant the quantities are said to vary in *inverse proportion*.

The graph is a curve like that on the right.

For example, the *time* you take to walk 12 km and the *steady speed* at which you walk vary in inverse proportion.

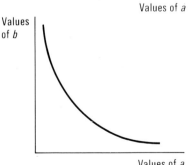

WITH A FRIEND

5   Decide between you which of these quantities vary in *direct* proportion, which vary in *inverse* proportion, and which vary in neither of these ways.

a) The *amount* of birthday cake available per person and the *number* of people at a party.

b) The *weight* of tomatoes you buy and their total *cost*.

c) The *time* for a telephone call and the *cost*.

d) The *rate of pay* and the *total earnings* for ten hours' work.

e) The *rate of pay* and the *number* of hours' work needed to earn £70.

f) The *volume* of milk in a jug and the *height* of the surface of the milk above the bottom of the jug.

6   Which tables represent two quantities which vary

(i) in direct proportion   (ii) in inverse proportion? (Sketching graphs will help.)

a)

p	0	1	2	3	4	5
q	2	3	4	5	6	7

b)

p	36	24	18	12	6	3
q	2	3	4	6	12	24

c)

p	0	1	2	3	4	5
q	0	1	4	6	16	25

d)

p	0	1	2	3	4
q	0	3	6	9	12

e)

p	36	24	18	12	6
q	3	2	1.5	1	0.5

f)

p	10	8	6	4	2	0
q	0	2	4	6	8	10

CHALLENGE

7   a) Two quantities a and b vary in direct proportion. a is 10 when b is 3. Find the value of b when a is 25.

b) Two quantities a and b vary in inverse proportion. a is 10 when b is 3. Find the value of b when a is 25.

ENRICHMENT

▨▨▨▨▨▨▨ EXPLORATION ▨▨▨▨▨▨▨

1 a) In the triangle the point X divides AB in the ratio 1 : 5.
The path shown in blue is produced by drawing lines parallel to the sides starting from X. In what ratio does the path divide each side each time it touches?

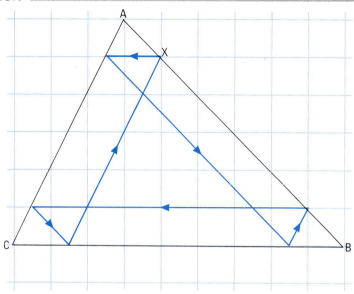

b) Draw a triangle of your own. Choose a point X along one side and write down the ratio in which it cuts the side. Draw a path like that in the diagram. In what ratio does the path divide each side each time it touches?

c) Investigate some more starting points. One path cuts the triangle into four identical smaller triangles. What is the ratio for this starting point?

d) Another path cuts the triangle into nine identical smaller triangles. What is the ratio for this starting point?

e) By choosing just two different starting points draw two paths which together cut the triangle into sixteen identical smaller triangles. What are the ratios for these starting points?

f) Investigate some more paths which cut the triangle into identical smaller triangles. Write a report about what you discover.

▨▨▨▨▨▨▨ ACTIVITY ▨▨▨▨▨▨▨

2 On a blank sheet of paper draw a large triangle with one side 20 cm long. Using a ruler and set square,

either ● divide your triangle into 400 smaller identical triangles
or ● divide it into any larger number you wish to choose.

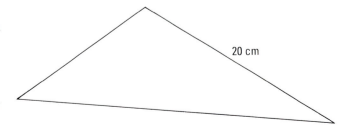

20 cm

B11 INDICES AND STANDARD FORM

REVIEW

We use indices to write repeated multiplications more simply, for example,

$10 \times 10 \times 10 \times 10 \times 10 \times 10 = 10^6$

$3.2 \times 3.2 \times 3.2 = 3.2^3$

■ Calculate   a) 2^3 b) 3^2.

10^{-3} means $\dfrac{1}{10^3}$, that is, $\dfrac{1}{1000}$

■ Calculate   (i) 3^{-2} (ii) 2^{-3}.

Any number to the power 0 is 1, for example, $10^0 = 1$, $2^0 = 1$, $3.7^0 = 1, \ldots$

CORE

1   The mean distance of the planet Venus from the Sun is 108 000 000 km. We can write this distance using indices in many ways, for example

108×10^6 km
or 1080×10^5 km
or 10.8×10^7 km, etc.

a) Write down three more examples of your own.

b) The mean distance of Mercury from the Sun is 57 900 000 km. Write the distance in three different ways using indices.

2   Write down which number in each pair is the larger:

a) 2.64×10^8, 2.6×10^8 b) 1.74×10^6, 1.74×10^5

c) 3.76×10^6, 37.6×10^6 d) 0.415×10^8, 41.5×10^5

3   Here are the mean distances of
the planets from the Sun:

Venus	10.8×10^7 km
Mars	228×10^6 km
Saturn	14.3×10^8 km
Earth	150×10^6 km
Jupiter	7.78×10^8 km
Mercury	579×10^5 km
Uranus	2870×10^6 km
Neptune	450×10^7 km
Pluto	5.9×10^9 km

a) Write each distance like this: $\square \times 10^{\square}$ km.

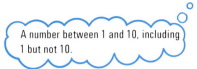

A number between 1 and 10, including 1 but not 10.

For example, Venus 1.08×10^8 km

b) Use your results to write the planets in order of distance from the Sun, starting with the one whose mean distance from the Sun is the least.

4   The amount of energy which a honeybee uses to beat its wings once is about 0.0008 joules.

Write this amount in the form $a \times 10^n$ where $1 \leqslant a < 10$ and n is a whole number. (Question 3 will help.)

───────────── TAKE NOTE ─────────────

Numbers written like this: 1.5×10^6
                               8.0×10^{-6}

are said to be in *standard form* or *scientific notation*.

A number in standard form is written in the form $a \times 10^n$, where $1 \leqslant a < 10$, and n is a whole number.

5   By writing each of these numbers in standard form, decide which is third largest:

$101.490\,14 \times 10^4$
$10\,149\,014$
0.1014×10^7
$0.001\,014\,9 \times 10^9$.

6   Write the number in each of these situations in standard form:

a) In a day's heavy manual labour you would use 17×10^6 joules of energy.

b) A burning match produces 4000 joules of energy.

c) The Earth receives 560×10^{22} joules of energy each year from solar radiation.

d) The mass of a carbon atom is about 0.2×10^{-22} g.

e) The Sun-grazing comet Ikeya-Seki passed within half a million miles of the Sun's surface in October 1965.

f) A cubic millimetre is $0.001 \times 0.001 \times 0.001 \text{ m}^3 = 0.000\,000\,001 \text{ m}^3$.

THINK IT THROUGH

7   a) Work out $40\,000 \times 30\,000$ in your head and write down the result.

b) Now do the calculation on a calculator, and copy the display.

c) Write one or two sentences to explain how a calculator records the results of 'large' calculations.

8   a) On a calculator press **1** **.** **2** **×** **1** **0** **x^y** **8** **=**

Write down the result of the calculation as a whole number.

b) How does a calculator record 2.6×10^9? Copy the display.

c) Use a calculator to find the results of these calculations. Write each result in standard form, $a \times 10^n$, where a is correct to 1 DP.

(i) $(6.2)^{14}$ (ii) $(2.3)^6 \times (1.7)^9$ (iii) 4×10^{-8}.

d) What keys must you press on your calculator to obtain this result? (That is, $0.000\,000\,001$.)

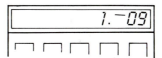

CHALLENGE

9   a) Astronomers deal with extremely large distances, so they use an extremely large unit of length – a *light-year*. Light travels at 300 000 km/s. A light-year is the distance that light travels in a year.
Use your calculator to find, as accurately as you can, how long a light year is in kilometres. Write your result in standard form.

b) The Earth's speed as it orbits the Sun is about 107 220 km/h. If the Earth could be used as a spaceship to travel from star to star, how many years would it take to reach the closest star, Proxima Centauri, 4.22 light-years away?

Reciprocals

▓▓▓▓▓▓▓▓▓▓▓▓▓▓▓ TAKE NOTE ▓▓▓▓▓▓▓▓▓▓▓▓▓▓▓

The reciprocal of 2 is 2^{-1}, that is, $\frac{1}{2}$.
The reciprocal of 3 is 3^{-1}, that is, $\frac{1}{3}$.
The reciprocal of $\frac{1}{2}$ is $\frac{1}{2}^{-1}$, that is, $\frac{1}{\frac{1}{2}}$ or 2.
The reciprocal of a is a^{-1}, that is, $\frac{1}{a}$.

1   What is the reciprocal of   a) 4   b) 1   c) $\frac{1}{6}$?

2   How many whole numbers have whole numbers as their reciprocals?

3   a) What is $1 \div \frac{1}{3}$?

   b) Use your result in a) to explain why 3 is the reciprocal of $\frac{1}{3}$.

   c) k is the reciprocal of m. Is it true or false that m must be the reciprocal of k? If you say 'false', give an example to explain your decision.

4   a) Write $1 \div \frac{5}{2}$ as a fraction.

   b) Explain why $\frac{2}{5}$ is the reciprocal of $2\frac{1}{2}$.

5   a) Write $1 \div \dfrac{a}{b}$ as a fraction.

   b) Explain why the reciprocal of $\dfrac{a}{b}$ is $\dfrac{b}{a}$.

6   What is the reciprocal of the reciprocal of:

   a) 2   b) $\frac{1}{4}$ c) $\frac{p}{5}$?

7   Write these as whole numbers, or vulgar fractions ($\frac{a}{b}$) or mixed numbers (for example, $2\frac{1}{2}$):

   a) the reciprocal of 0.2

   b) the reciprocal of 0.8

   c) the reciprocal of 1.6.

8   a) The reciprocal of a number is 0.2. What is the number?

   b) Find the fraction whose reciprocal is 0.6.

9   a is the reciprocal of b. What is:

   a) ab b) $\dfrac{a}{b}$ c) $\dfrac{b}{a}$?

   Write each result (i) in terms of a only (ii) in terms of b only.

━━━━━━━━━━ CHALLENGE ━━━━━━━━━━

10 The reciprocal of $\dfrac{x-y}{x}$ is 10.

a) Find a possible pair of values for x and y.

b) How many possible pairs are there? If there is more than one, describe the set of possible pairs.

ENRICHMENT

1 a) Write: (i) $5^8 \times 5^4 \times 5^2$ in the form $5^\square$.
        (ii) $3.7^2 \times 3.7^4 \times 3.7$ in the form $3.7^\square$.

b) Try some more examples of your own like those in a). Find a quick way of arriving at the result of a multiplication such as $12^{24} \times 12^{13}$.

c) Investigate some divisions, for example, $5^4 \div 5^3 = 5$ $\qquad$ $12^4 \div 12^9 = 12^{-5}$.

Find a quick way of arriving at the result of a division such as $9^9 \div 9^{17}$.

2 Use your results in question 1 to do these in your head (write each result in the form a^x):

a) $2.1^4 \times 2.1^7$ $\qquad$ b) $18^8 \times 18^{-3}$ $\qquad$ c) $5^{-4} \times 5^{-12}$

d) $\dfrac{3^{19}}{3^7}$ $\qquad$ e) $2^{-3} \div 2^{-3}$ $\qquad$ f) $5^8 \div 5^{-6}$.

3 a) Estimate the value of: (i) $4^{1.5}$ (ii) $4^{0.5}$.

b) Multiply your results in a) together to give an approximate result for $4^{1.5} \times 4^{0.5}$.

c) Copy and complete: $4^{1.5} \times 4^{0.5} = 4^\square = \square$.
Compare your result with your result in b).
How close do you think your estimates were in a)? (Close, very close, or not very close.)
Check each one with your calculator.

CORE

--- TAKE NOTE ---

The tiles in this pattern have identical shapes and sizes, that is, they are all exactly the same shape and size. We say they are *congruent*.

--- ACTIVITY ---

1 You need a ruler and a protractor.
Measure lengths and angles. Decide which shape, A or B is congruent to the blue shape. For each shape that is *not* congruent, explain why.

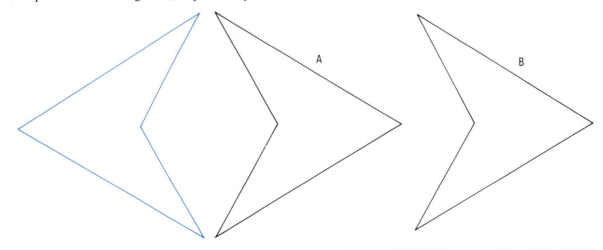

2   This tiling pattern is made
    from two different sizes of
    triangle:

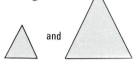 and

a) The larger triangles are
   enlargements of the
   smaller triangles. What
   is the scale factor for
   enlargement?

b) These are two of the
   small triangles.

(i) Are they congruent?   (ii) Are they similar?   (iii) What is the scale factor for enlargement?

ACTIVITY

c) Sketch a tiling pattern of your own that is made of two different sets of similar shapes.

B12

ACTIVITY

3 a) You need a ruler and a pair of compasses.
      This triangle is drawn full size. Draw a triangle
      that is congruent to it.

10 cm   7 cm

15 cm

b) Draw a triangle congruent to triangle A
   (not drawn to scale).

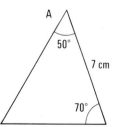

c) Draw a triangle which is *not* congruent to
   triangle A but which has

        ● a 7 cm side
   and ● 50° and 70° angles.

A

50°

7 cm

70°

CHALLENGE

4   A triangle has two of its sides 6 cm and 10 cm long. One of its angles is 40°. Another triangle also has two of its sides 6 cm and 10 cm long, and an angle of 40°.
But the two triangles are *not* congruent. Draw each of them, full size.

5   a)   Do these produce *congruent* or *similar* (but not congruent) shapes?

A Slide projector   B A jelly mould   C A date stamp

b)   Name something else which produces (i) congruent shapes (ii) similar (*but not congruent*) shapes.

TAKE NOTE

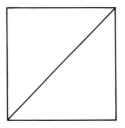   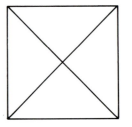   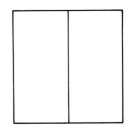

Congruent   *Similar*

6   A square can be made from:

two congruent triangles   or   four congruent triangles   or   two congruent rectangles.

a)

Any parallelogram can be divided into four congruent triangles.
Helen

That's not true.
Glyn

Who is correct, Helen or Glyn? If you say Glyn, sketch a parallelogram to explain.

b) A regular hexagon can be made from two pairs of different congruent triangles. Make a sketch to show how.

c) Every kite can be made from two pairs of different congruent triangles and two congruent parallelograms (that is, six shapes altogether). Make sketches to show how.

━━━━━━━━ WITH A FRIEND ━━━━━━━━

7   It is possible to draw two quadrilaterals whose sides are identical in length, but which are not congruent. For example,

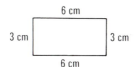 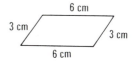

Decide between you if it is possible to draw two different (that is, non-congruent) figures in cases A to J. If you decide it is possible to draw two *different* figures, do so.

A   Two quadrilaterals with the same sized angles.

B   Two quadrilaterals with the same sized angles and the same length sides.

C   Two triangles with two pairs of angles identical to each other.

D   Two triangles with two pairs of sides identical to each other.

F   Two triangles with one pair of sides identical.

E   Two triangles with one pair of angles identical.

H   Two triangles with a pair of identical angles and two pairs of identical sides.

G   Two triangles with a pair of identical angles and a pair of identical sides.

J   Two triangles with two pairs of identical angles and a pair of identical sides.

I   Two triangles with three pairs of identical sides.

━━━━━━━━ TAKE NOTE ━━━━━━━━

Triangles whose sides are the same length must be congruent.

Triangles with two sides of the one equal to two sides of the other, and the angle between them equal, must be congruent.

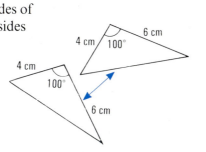

Triangles with two equal sized angles and one pair of *corresponding* sides of equal length must be congruent.

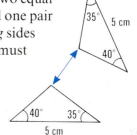

■■■■■■■■■■■■ ASSIGNMENT ■■■■■■■■■■■■

8   Think of quadrilaterals with four sides all 2 cm long.

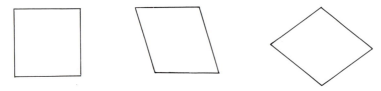

   a)   You are told that two of the shapes also have an equal angle. Does this mean that the two shapes are congruent?

   b)   If you say 'No' in a), how many angles must be identical before we can be sure that the shapes are congruent?

   c)   Now think about quadrilaterals whose angles are right angles. You are told that two of the shapes have one pair of sides the same length. Does this mean that the shapes are congruent?

   d)   If you say 'No' in c), how many pairs of sides must be identical before we can be sure that the shapes are congruent?

   e)   Investigate convex kites. What is the minimum we must know to be the same about two of them, before we can be sure that they are congruent?

   f)   Investigate some more quadrilaterals, such as squares, rectangles, parallelograms, trapeziums, . . . What is the least we must know about two of each type before we can be sure that they are congruent?

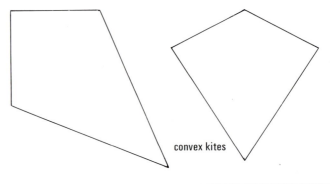

convex kites

B13 SCATTER GRAPHS

1 In mathematics we draw graphs to show how two quantities relate to each other. For example,

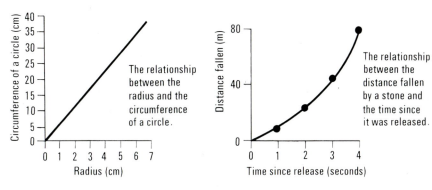

The relationship between the radius and the circumference of a circle.

The relationship between the distance fallen by a stone and the time since it was released.

Often, however, quantities don't relate to each other as neatly as this. Here is an example.

Ten families were asked how many brothers and sisters there were in the family. The results are shown here, written *in a table* and *on a graph* (called a scatter graph).

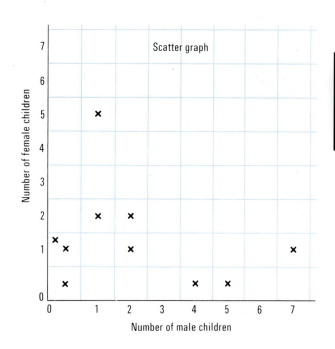

Number of male children	Number of female children
2	1
5	0
2	2
0	1
1	2
0	1
7	1
1	5
4	0
0	0

Below are the results for ten more families. Copy the scatter graph and add these results to it.

Number of male children	2	0	1	2	1	1	1	3	3	1
Number of female children	2	2	1	4	6	0	0	0	1	1

2   a)   The scatter graph shows the heights and weights of the five
dogs in the photographs. Try to match each point on the
graph with a dog.

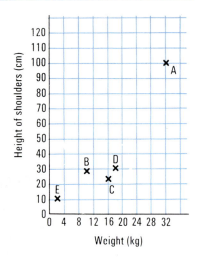

A   Boxer

B   Greyhound

C   Great Dane

D   Beagle

E   Yorkshire Terrier

b)   Copy the scatter graph. Mark points F and G for two more dogs (you will need
to estimate their heights and weights).

3   This is the scatter graph for the dogs in question 2.

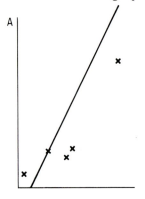
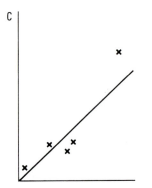
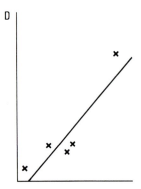

a)   Although the points do not all lie neatly on a line, we can draw a line by eye which
seems to fit best. Here are some attempts. Which do you think fits best?

b) Graph C is probably the 'best fit'. (It also suggests that a dog which weighs 0 kg has height 0 cm, whereas the others do not.) Use graph C to estimate the weights and heights of these dogs:

(i) Dalmatian: height 50 cm, weight ? kg

(ii) Bulldog: height ? cm, weight 24 kg

(iii) St Bernard: height 65 cm, weight ? kg

4   A is a scatter graph of the distance fallen by a tennis ball and its height of bounce. B is a scatter graph of the best times for an athlete over various distances.

a)   Draw each scatter graph on 1 cm squared paper and estimate the line of 'best fit'. (Decide first if you think the line should be straight.)

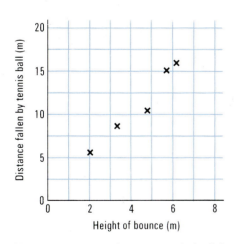

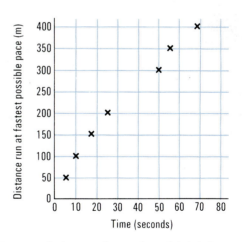

b)   Suppose you drop a tennis ball from 25 m. Use graph A to estimate how high it bounces.

c)   Use graph B to estimate the athlete's best time for a distance of 250 m.

▓▓▓▓▓▓▓▓▓▓▓▓ TAKE NOTE ▓▓▓▓▓▓▓▓▓▓▓▓

A scatter graph indicates how two quantities are related. Sometimes a trend can be detected and we can draw a line of 'best fit'. We can use the line of 'best fit' to help us to make predictions for measurements we have not taken.

■■■■■■■■■■■■■■■■■■ ACTIVITY ■■■■■■■■■■■■■■■■■■

5 a) You need some plain paper.
       On the paper you are going to sketch a set of 10 rectangles with lengths
       1 cm, 2 cm, ... , 9 cm, 10 cm. As you draw each rectangle, choose its width
       so that you produce the shape of rectangle that you find most pleasing.

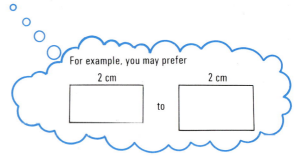

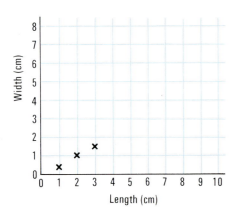

   b) Draw a scatter graph of your results.

   c) Is there a definite trend in your results? If so, draw a line of 'best fit'.

   d) Roughly what ratio of length to width of a rectangle do you appear to find most pleasing?

ENRICHMENT

■■■■■■■■■■■■■■■■■■ ASSIGNMENT ■■■■■■■■■■■■■■■■■■

1   Think of two variables (such as heights and weights of people, number of rooms and number of
    windows in houses, ages of people and distances they travel to work, number of petals and
    number of leaves on buttercups, ...). Design a survey to collect information which you can
    represent on a scatter graph.

    Draw a scatter graph and decide if there is a line of 'best fit'. Write a report to explain what you
    did and what your results suggest.

2   The table gives the heights, waist measurements, and head sizes for a group of 30 teenagers. Which do you think are most clearly related,

- heights and waist measurements
or - heights and head sizes
or - waist measurements and head sizes?

Write a report to explain how you decided and what conclusions you reached.

Height (cm)	Waist (cm)	Head size (cm)
159	68	17.0
163	68	17.5
165	69	18.0
169	72	19.0
169	70	16.5
172	75	17.5
174	69	19.0
174	73	20.5
174	78	20.5
175	76	18.0
176	68	17.5
177	69	16.5
180	70	21.0
181	77	17.5
183	74	17.0
184	80	18.5
185	75	21.5
187	79	20.5
187	82	18.0
188	77	17.5
189	74	18.5
190	79	20.5
191	72	21.0
193	80	19.5
193	84	19.0
194	73	18.5
194	77	17.5
195	77	18.0
196	80	18.0

B13

REVIEW

- We use brackets to show which pairs of numbers are combined. For example,

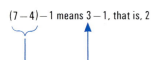 $(7-4)-1$ means $3-1$, that is, 2

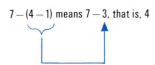 $7-(4-1)$ means $7-3$, that is, 4

■ Use one pair of brackets to make this equal to 5: $10-8-2+1$.

- To show that two expressions are simply different ways of writing the same thing we use '$\equiv$'.

  So, for example: $a+a+b \equiv 2a+b$, and $a(b+c) \equiv ab+ac$.

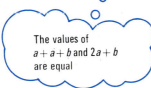
The values of $a+a+b$ and $2a+b$ are equal

We say 'identically equal to'

We call expressions like $a+a+b \equiv 2a+b$ *identities*.

■ What is the difference between an *equation* and an *identity*? Give an example of each to help you to explain the difference.

CORE

 EXPLORATION

1   Sometimes the positioning of brackets changes the result of a calculation, for example,

$9-(4+1)$ is not equal to $(9-4)+1$

    4   6

Sometimes the positioning of brackets does *not* change the result of a calculation:

$(9+4)+1$ is equal to $9+(4+1)$

    14   14

Investigate when it is important to use brackets and when it is not.
Try different combinations of the four operations $+, -, \times, \div$.
Write a report about what you discover.

2 a, b and c are variables. Each one can represent any number. Write down which of these are true and which are false.

a) $a+(b+c)\equiv(a+b)+c$ b) $a-(b-c)\equiv(a-b)-c$

c) $a\times(b\times c)\equiv(a\times b)\times c$ d) $a\div(b\div c)\equiv(a\div b)\div c$

e) $a\times(b+c)\equiv(a\times b)+c$ f) $(a-b)\times c\equiv a-(b\times c)$

g) $\dfrac{a}{b+c}\equiv\dfrac{a}{b}+\dfrac{a}{c}$ h) $\dfrac{a+b}{c}\equiv\dfrac{a}{c}+\dfrac{b}{c}$

i) $\dfrac{a}{b-c}\equiv\dfrac{a}{b}-\dfrac{a}{c}$ j) $\dfrac{a-b}{c}\equiv\dfrac{a}{c}-\dfrac{b}{c}.$

3 a) Calculate $8-(3-2)$ and $8-3+2$. (When there are no brackets, work from left to right – that is, subtract then add in this case.) Write down what you notice.

b) Calculate $12-(4-5)$ and $12-4+5$. Write down what you notice.

c) Try some examples of your own. Use your results to help you complete this:

$a-(b-c) = a?b?c$

4 a) Calculate $8-(3+2)$ and $8-3-2$. Write down what you notice.

b) Calculate $9-(1+6)$ and $9-1-6$. Write down what you notice.

c) Try some examples of your own. Use your results to help you to complete this:

$a-(b+c) = a?b?c$

5 Try different sets of numbers for a, b and c to help you to complete:

a) $a+(b+c)\equiv a?b?c$

b) $a+(b-c)\equiv a?b?c$

TAKE NOTE

When there is a $+$ outside of a bracket, and the bracket is *removed*, the sign $+$ or $-$ inside the bracket does *not* change.

$p+(q+r)\equiv p+q+r$
$p+(q-r)\equiv p+q-r$

When there is a $-$ outside of a bracket, and the bracket is *removed*, the sign inside the bracket *changes* from $+$ to $-$, or $-$ to $+$.

$p-(q+r)\equiv p-q-r$
$p-(q-r)\equiv p-q+r$

---

6 Some of these are correct and some are incorrect. Correct the ones which are incorrect.

a) $2+(4-3) = 2+4+3$ b) $8-(4-3) = 8-4+3$ c) $5-(2+1) = 5-2+1$

d) $8+(7-5) = 8+7-5$ e) $a-(1+b) = a-1-b$ f) $a+(1-c) = a+1-c.$

B14

7   Write each of these without brackets.
    Simplify each of your results as much as possible.

> For example, $5-(k+2)$
> $\equiv 5-k-2$
> $\equiv 3-k$

a) $n-(m+t)$ b) $5-(3-c)$ c) $4-(t+1)$ d) $n-(10+n)$

e) $2k-(5+n)$ f) $3n-2(4+n)$ g) $p+2(p-1)$ h) $5k-3(k+2)$

i) $3t-3(4+t)$ j) $(m-1)-(m-2)$ k) $3(m-1)-2(m-1)$.

> $3n-(8+2n)$
> $\equiv \ldots$

8   Simplify each expression:

a) $2(a-b)-2(c-a)+2(b-c)$ b) $2(k+n)+(k+n)-(k-n)$ c) $(3n-2)-(2-3n)-(n-3)$.

Inserting brackets

> Remember?
> When there are no brackets we carry
> out each operation in order, from left to right.

1   a) Check that $19+7-4=22$.

b) Start with $19+7-4$. Insert brackets: $19+(7-4)$. Has the result changed?

c) Start with $19-7+4$. Insert brackets: $19-(7+4)$. Has the result changed?

d) Start with $19+7+4$. Insert brackets: $19+(7+4)$. Has the result changed?

e) Start with $19-7-4$. Insert brackets: $19-(7-4)$. Has the result changed?

f) Choose different sets of numbers for a, b and c to help you to complete:

    (i) $a+b+c\equiv a?(b?c)$ (ii) $a+b-c\equiv a?(b?c)$
    (iii) $a-b+c\equiv a?(b?c)$ (iv) $a-b-c\equiv a?(b?c)$.

━━━━━━━━━━━━━━ TAKE NOTE ━━━━━━━━━━━━━━

When we *insert* brackets after a $+$ sign, the $+$ or $-$ sign which
follows does *not* change inside the brackets.

$$p+q+r\equiv p+(q+r)$$
$$p+q-r\equiv p+(q-r)$$

When we *insert* brackets after a $-$ sign, the $+$ or $-$ sign which
follows *changes* to $-$ or $+$ inside the brackets.

$$p-q+r\equiv p-(q-r)$$
$$p-q-r\equiv p-(q+r)$$

B14

6   Insert brackets after the first $-$ sign in each of these expressions. You may need to change some
    signs to make sure that the result is equivalent to the original.

a) $5-4-1$ b) $8-7+6$ c) $p-q+r$ d) $n-t-q$ e) $t-2q-2r$

f) $p-3n+3t$ g) $t-3k+6$ h) $m-4n-8$ i) $m-6-2t$

j) $5-2t+4k$ k) $v-12+6k$ l) $m-5t-50$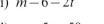

m) $2t-5n+20$ n) $5-kx+x$ o) $10-x+kx$ p) $8-xy-x$

> $t-3(k\ldots)$

q) $8-xy-y$ r) $5-2k+tk$ s) $m-5n-xn$.

> $5-x(k\ldots)$

ENRICHMENT

Brackets within brackets

1   Remove the brackets in these expressions. Simplify your results as much as possible.

   a) $3(n-p)-3(p-n)-3(n+p)$

   b) $a-[b-(a-b)]$ (Always deal with the inner brackets first.)

   c) $a-b[a-b(a-b)]$.

2   Copy and complete: $a-b-c-d\equiv(a?b)-(c?d)$

$$\equiv a-[b?(c?d)].$$

3   Calculate:

   a) $2-(3+{}^-4)$ b) $5-({}^-3-{}^-4)$.

4   a) Which is correct: $4-{}^-2+3$ is   (i) $4-({}^-2+3)$
                                  or (ii) $4-(2-3)$
                                  or (iii) $4-({}^-2-3)$?

   b) Which is correct: $4-{}^-2-3$ is   (i) $4-({}^-2-3)$
                                  or (ii) $4-(2+3)$
                                  or (iii) $4-({}^-2+3)$?

5   Simplify, as much as possible:

   a) $4-(a-3)-2(a-5)+2[3-(a-1)]$ b) $p-(q-r)+3(r-q-1)-2[2q-(r+3)]$.

6   Solve the equations and check that your solution to each one is correct:

   a) $4[p-(2-p)]=16$ b) $3[1-(5+p)]=p-20$.

REVIEW

The part of mathematics which deals with angles and lengths is called *trigonometry*.

- In a right-angled triangle like PQR we give some ratios special names:

$\dfrac{PQ}{RQ}$ is called tan $x°$. ∘∘∘

$\dfrac{PQ}{RP}$ is called sin $x°$. ∘∘∘

$\dfrac{RQ}{RP}$ is called cos $x°$. ∘∘∘∘

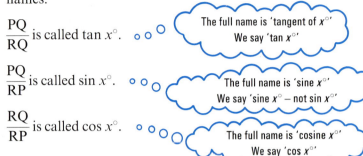

The full name is 'tangent of $x°$'
We say 'tan $x°$'

The full name is 'sine $x°$'
We say 'sine $x°$ – not sin $x°$'

The full name is 'cosine $x°$'
We say 'cos $x°$'

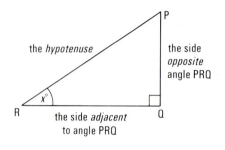

the *hypotenuse*

the side *opposite* angle PRQ

the side *adjacent* to angle PRQ

- You should know these facts: $\tan x° = \dfrac{\text{opposite}}{\text{adjacent}}$ $\sin x° = \dfrac{\text{opposite}}{\text{hypotenuse}}$ $\cos x° = \dfrac{\text{adjacent}}{\text{hypotenuse}}$

■ In triangle ABC, $\sin y° = \dfrac{AC}{AB}$. Write a similar ratio for

  a) $\sin z°$ b) $\cos z°$ c) $\tan z°$ d) $\cos y°$ e) $\tan y°$.

- For triangle LMN, when we press

  [C] [4] [3] [tan] [=] ,

  the calculator displays the ratio $\dfrac{LM}{MN}$: | **0.932515** |

  So tan 43° = 0.93 (to 2 DP).

  [C] [4] [3] [sin] [=] , displays the ratio $\dfrac{LM}{LN}$.

  [C] [4] [3] [cos] [=] , displays the ratio $\dfrac{MN}{LN}$.

■ Find (i) $\sin \angle MLN$ (ii) $\cos \angle MLN$ (iii) $\tan \angle MLN$.

- We can find the distance AB, like this:

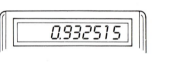

$\sin 38° = \dfrac{4}{AB} = \dfrac{4}{k}$

$\times k: \ k \sin 38° = 4$

$\div \sin 38°:$

$k = \dfrac{4}{\sin 38°}$

$= 6\cdot5 \, (\text{1 DP})$

- Pythagoras's rule tells us that the area of A is the sum of the areas of B and C.

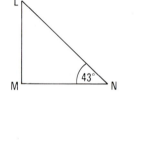

■ Find AC by using tan. Write your result correct to 1 DP.

CONSOLIDATION

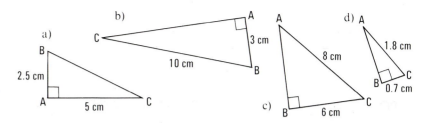

1   Use Pythagoras's rule to find the length of the third side of each triangle. Write your results correct to 1 DP.

2   Find the length of the side marked '*x* cm' in each triangle. Write each result correct to 1 DP.

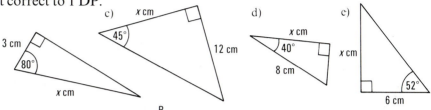

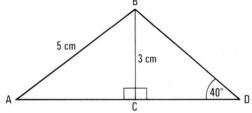

3   a) Use Pythagoras's rule to find AC.

   b) Use sin, cos or tan to find (i) BD (ii) CD.

   c) How long is AD?

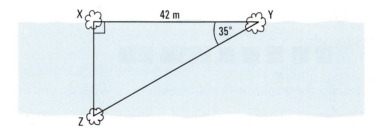

4   X, Y and Z are three trees on the banks of a river. Angle ZXY is 90° and XY is 42 m. Angle XYZ = 35°. Roughly, how wide is the river at point X?

5   PQR is an equilateral triangle, and PQ = 8 cm. What is the height, PX, of the triangle?

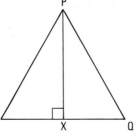

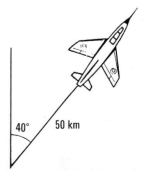

6   The aeroplane is flying away from the airport on a bearing of 040°.

   a) How far West of the airport is the aeroplane when it has travelled 50 km?

   b) How far North of the airport is it?

B1

Finding angles

1   a) Use your calculator to find the angle whose tangent is 0.24.
For example, try

| C | 3 | 0 | tan |

| C | 2 | 0 | tan | and so on.

Find the angle to the nearest degree.

b) Suppose we want to find the size of angle CAB in this triangle.

We know that $\tan x° = \dfrac{24}{100} = 0.24$.

Use your result in a) to write down $x°$ to the nearest degree.

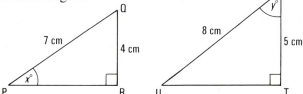

c) Use the method in parts a) and b) to find
$x°$ and $y°$ in triangles PQR and STU,
correct to the nearest degree.

d) Press | C | . | 2 | 4 | INV | tan |

Compare the display with your result in a). What do you think the display tells you?

e) Press | C | 4 | ÷ | 7 | = | INV | sin |

Compare the display with your results in c). What do you think the display tells you?

f) Use your calculator to find the value of $y°$ in triangle STU, correct to 2 DP.

g) The tangent of an angle is 2.82. Use the | INV | and | tan | keys to find the size of
the angle correct to 2 DP.

2   a) Find $\sin x°$ for each of these triangles:

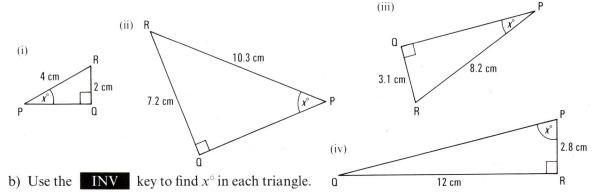

b) Use the | INV | key to find $x°$ in each triangle.

3   a) Find cos $x°$ for each of these triangles:

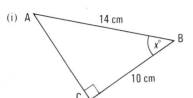   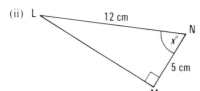   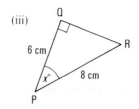

   b) Use the ▋INV▋ key to find $x°$ in each triangle correct to 1 DP.

4   a) Find tan $x°$ for each of these triangles:

(iii)

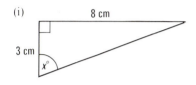   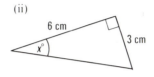   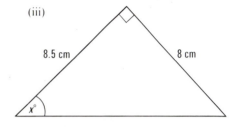

   b) Use the ▋INV▋ key to find $x°$ in each triangle correct to 1 DP.

───── TAKE NOTE ─────

▋C▋ ▋.▋ ▋2▋ ▋4▋ ▋INV▋ ▋tan▋ gives the angle whose tangent is 0.24, that is, 13.5° (1 DP).

▋C▋ ▋.▋ ▋3▋ ▋9▋ ▋INV▋ ▋cos▋ gives the angle whose cosine is 0.39, that is, 67.0° (1 DP).

▋C▋ ▋.▋ ▋4▋ ▋2▋ ▋INV▋ ▋sin▋ gives the angle whose sine is 0.42, that is, 24.8° (1 DP).

5   Find $p°$ in each triangle correct to 1 DP.

a)   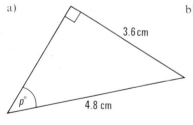   b)   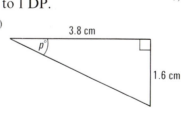   c)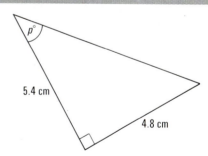

6   Imagine that you draw a diagonal line across this page from corner to corner. Calculate, to the nearest degree, the angle between the diagonal and the longer edge. Check your calculation by measuring with a protractor.

B15

───── CHALLENGE ─────

7   You need 1 cm squared paper.
   You are allowed to use only a ruler and a calculator. You cannot use a protractor. Draw a right-angled triangle which has one angle of 42°. Explain how you solved the problem.

8   a) This is a full-sized design drawing for a new stamp. The
       angle at the top is 90°. Find the sizes of the other two
       angles in each of these ways (you can check the results
       against each other):

       (i)   by estimating by eye
       (ii) by measuring with a protractor
       (iii) by measuring the lengths of sides and then using a calculator

    b) How accurate was   (i) your estimation by eye,
                           (ii) your measurement with the protractor?

9   On this map 1 cm represents 1 km.
    Find these bearings to the nearest degree:

    a) Farlow from Hollingway

    b) Edgerton from Hollingway

    c) Farlow from Edgerton.

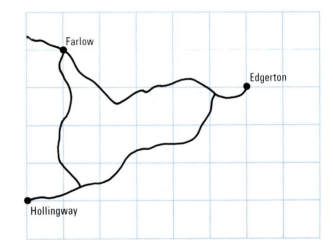

10 In the rectangle ABCD, find

    a) $x°$ b) $y°$.

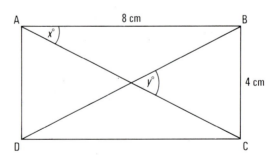

11 In the quadrilateral ABCD, find

    a) AC   b) ∠DAC.

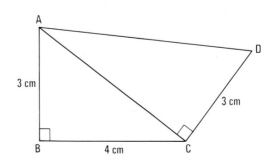

════════════════════ CHALLENGE ════════════════════

12 A ship sails away from a harbour for 20 km in a direction
025°, and then due East for 15 km.

a) How far   (i) North of the harbour
             (ii) East of the harbour
   is the ship at the end of its journey?

b) Another ship sets out from the same
   harbour to travel directly to the
   same point. In which direction
   should it travel?

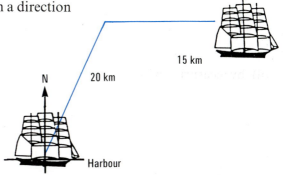

ENRICHMENT

1   In the diagram find
    a) AC b) AD. Write each
    result correct to 1DP.

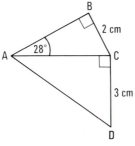

2   a) Use the diagram of triangle ABC to help you to
       explain why the value of tan 45° must be 1.

    b) Suppose AB = 1 cm. Explain why $\sin 45° = \dfrac{1}{\sqrt{2}}$.

    c) What is cos 45°?

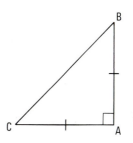

3   a) Use the diagram of triangle LMN to explain why sin 50°
       and cos 40° are equal.

    b) Copy and complete:

       (i) $\sin 25° = \cos \square °$ (ii) $\cos 37° = \sin \square °$

       Check each of your results with a calculator.

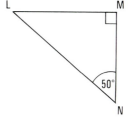

4   a) You need 1 mm squared paper.

       Copy the axes of the chart. Draw the
       graph of (i) $y = \sin x°$ (ii) $y = \cos x°$.

    b) From your graph find the value of $x°$ for
       which $\sin x° = \cos x°$. Check your result
       with a calculator.

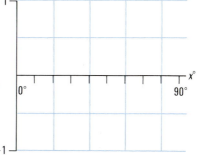

B15

5 a) In triangle PQR find a
       formula for the distance
       PX using

       (i) $\sin A°$ and b cm
       (ii) $\sin B°$ and a cm.

   b) Explain why

       $$\frac{a}{b} = \frac{\sin A°}{\sin B°}.$$

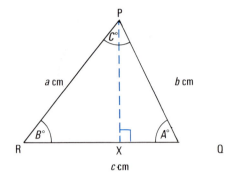

CHALLENGE

   c) Make another copy of the triangle PQR and draw a perpendicular from Q to PR. Use your
       diagram to explain why

       $$\frac{b}{c} = \frac{\sin B°}{\sin C°}.$$

B15

REVIEW

● Equations like $x + y = 10$

have an infinite number of solutions. Here are three solutions of $x + y = 10$:

$x = 1, y = 9$; $x = 0.5, y = 9.5$; $x = {}^-1, y = 11$.

■ Write down three solutions of $2x + y = 4$.

● We can represent the solutions of an
equation as a line on a graph.
Here is the graph for the solutions of
$2x + y = 4$:

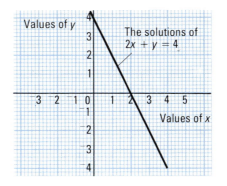

● When we search for values of x and y
which satisfy two (or more) equations, we
say we are solving *simultaneous equations*.

There is just one solution of $x + y = 6$
that is also a solution of $y = x - 2$, namely
$x = 4, y = 2$.

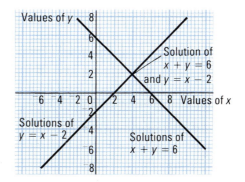

■ Do the simultaneous equations $x + y = 6$,
$$y = x - 1$$
have a solution? If so, what is it?

(Guess and test values of x and y.)

CONSOLIDATION

1   Copy and complete these solutions of the equation $2y + 1 = x$:
   $x = 1, y = \square$; $x = 3, y = \square$; $x = \square, y = 5$; $y = 1.3, x = \square$.

2   Sketch lines to represent the solutions of each of these equations:

   a) $x + y = 4$ b) $x - y = 4$ c) $4y = x$.

Finding equations and their solutions

TAKE NOTE

On a grid we call the horizontal axis the *x axis* and the vertical axis the *y axis*. We call the first coordinate of any point the *x coordinate* and the second coordinate the *y coordinate*.

$(2, {}^-1)$

x coordinate   *y* coordinate

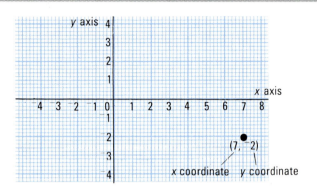

---

1   Previously we have been given equations, or we have represented problem situations by an equation, and then we have found the solutions. We have seen that solutions to equations with two letters can be represented by a straight line or a curve on a grid. Now we will find the equations of straight lines.

a) Check that points on line A are solutions of the equation $y = x + 1$.

We say that $y = x + 1$ is an equation of line A.

b) Check that $y = 1 - x$ is an equation of line B.

c) Check that $y = x - 5$ is an equation which has these values as solutions for x and y:

$(2, {}^-3) (5, 0) (7, 2) (9, 4)$

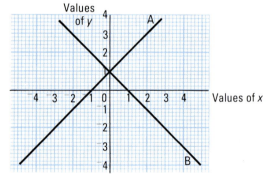

2   The table gives six points on a line in the (x, y) plane (a plane with x values along the horizontal axis and y values along the vertical axis):

x	$^-2$	$^-1$	0	1	2	3
y	$^-3$	$^-1$	1	3	5	7

a) Check that $y = 2x + 1$ is an equation of the line.

b) Write the equation in the form:
(i) $y... = 1$ (ii) $x = ...$

c) Draw the line on a grid.

3   The table shows seven points on a curve in the (x, y) plane:

x	$^-3$	$^-2$	$^-1$	0	1	2	3
y	10	5		1	2		

a) Check that $y = x^2 + 1$ is an equation of the curve.

b) Write down the three missing coordinates.

c) Write the equation in the form:
(i) $y... = 1$ (ii) $x = ...$

d) Draw the curve on a grid.

━━━━━━━━━━━━ CHALLENGE ━━━━━━━━━━━━

4 a) Find the equation of the line (or curve) for each table of values.
   b) Write each equation in another way.
   c) Draw each line (or curve) on the same set of axes.

(i)

x	⁻4	⁻3	⁻2	⁻1	0	1	2
y	⁻9	⁻7	⁻5	⁻3	⁻1	1	3

ii)

x	9	4	1	0	1	4	9
y	⁻3	⁻2	⁻1	0	1	2	3

iii)

x	⁻3	⁻2	⁻1	0	1	2
y	⁻2	⁻1	0	1	2	3

iv)

x	⁻4	⁻3	⁻2	⁻1	1	2	3	4
y	⁻3	⁻4	⁻6	⁻12	12	6	4	3

5 Match each equation with a line (or curve):

$$y - x = 0 \qquad y - 3 = x \qquad y + x + 3 = 0 \qquad y^2 = x + 1 \qquad xy = 24 \qquad x^2 = y + 1.$$

a)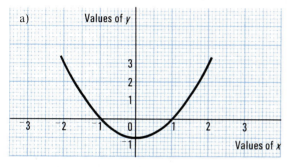

b)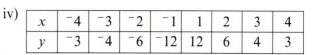

c)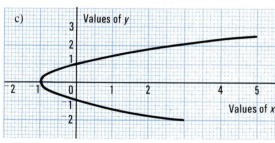

d)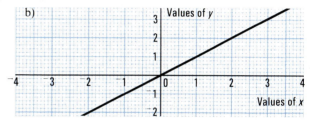

e)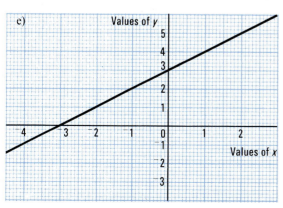

f)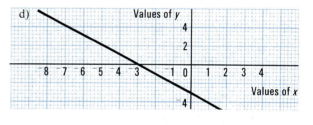

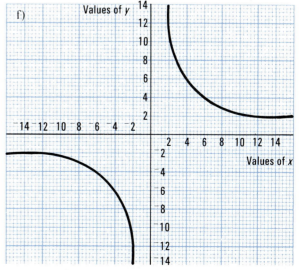

B16

Coordinates

1   In this set of coordinates, check that the x and y coordinates in each pair obey the rule $y = x + 1$:

$(^-3, \ ^-2)\,(^-2, \ ^-1)\,(^-1, 0)\,(0, 1)\,(1, 2)\,(2, 3)\,(3, 4)\,(4, 5)$

Write down a set of eight coordinates which obey the rule $y = 3x - 4$.

2   In this set of coordinates only the x coordinate is given.

$(^-4, \quad)\,(^-3, \quad)\,(^-2, \quad)\,(^-1, \quad)\,(0, \quad)\,(1, \quad)\,(2, \quad)\,(3, \quad)\,(4, \quad)$

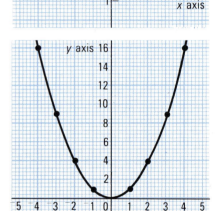

a)   Many different choices can be made for the y coordinates. Here are two sets of choices and a sketch of the graphs they produce (each graph assumes there are many points between the ones given):

A $(^-4, 3)\,(^-3, 2)\,(^-2, 1)\,(^-1, 0)\,(0, 0)\,(1, 1)\,(2, 2)\,(3, 1)$ $(4, 0)$

B $(^-4, 16)\,(^-3, 9)\,(^-2, 4)\,(^-1, 1)\,(0, 0)\,(1, 1)\,(2, 4)\,(3, 9)$ $(4, 16)$

In one of the sets a simple rule connects x and y in each pair of coordinates. Which set is it? What is the rule? (Write it as $y = \ldots$)

b)   Write down your own y coordinates for the set so that:

(i)   no simple rule connects x and y in each pair of coordinates

(ii) a simple rule does connect the x and y coordinates.

Draw a graph for each of your sets.

c)   Write down the y coordinates for the set so that the x and y coordinates obey the rule $y = 2x + 1$. Draw the graph for the rule.

d)   (i)   Write down the y coordinates for the set so that the graph they produce is a straight line passing through $(2, 4)$ and $(0, 2)$.
     (ii) Is there a simple rule which connects the x and y coordinates? If so, write it down.

—————— EXPLORATION ——————

3   Complete the set of coordinates $(^-4, \quad)\,(^-3, \quad)\,(^-2, \quad)\,(^-1, \quad)\,(0, \quad)\,(1, \quad)\,(2, \quad)\,(3, \quad)\,(4, \quad)$ in different ways to produce different graphs such as these:

Investigate which shapes of graphs can be represented by simple rules connecting x and y. Write a report about what you discover.

B16

4   Find the rule connecting x and y for each set of coordinates:

   a) $(^-4, ^-4)$ $(^-2.5, ^-2.5)$ $(0, 0)$ $(3, 3)$ $(5.2, 5.2)$ $(7, 7)$...

   b) $(^-2, ^-4)$ $(^-1, ^-3)$ $(0, ^-2)$ $(1, ^-1)$ $(2, 0)$ $(5.1, 3.1)$...

   c) $(^-6, 3)$ $(^-4, 2)$ $(^-1, \frac{1}{2})$ $(0, 0)$ $(2, ^-1)$ $(3, ^-1\frac{1}{2})$ $(6, ^-3)$...

   d) $(24, 1)$ $(12, 2)$ $(6, 4)$ $(8, 3)$ $(3, 8)$ $(4, 6)$ $(2, 12)$ $(1, 24)$...

   e) $(2, ^-4)$ $(3, ^-9)$ $(5, ^-25)$ $(6, ^-36)$ $(9, ^-81)$...

   f) $(^-3, ^-3)$ $(^-2, ^-4)$ $(^-1, ^-5)$ $(0, ^-6)$ $(1, ^-7)$ $(2, ^-8)$...

CHALLENGE

5   Find the rule connecting x and y for each set of coordinates:

   a) $(^-4, ^-16)\,(^-3, ^-13)\,(^-2, ^-10)\,(^-1, ^-7)\,(0, ^-4)\,(1, ^-1)\,(2, 2)$...

   b) $(^-3, 11)\,(^-2, 6)\,(^-1, 3)\,(0, 2)\,(1, 3)\,(2, 6)\,(3, 11)$...

TAKE NOTE

The coordinates on the curve satisfy the equation

$$y + x^2 = 0$$

We say that $y + x^2 = 0$ is the equation of the curve.

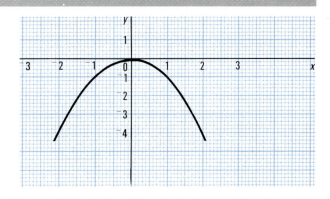

EXPLORATION

6   Think of some straight lines through point A on the graph. Find their equations.

   Try some different starting points.
   Write a report about what you discover.

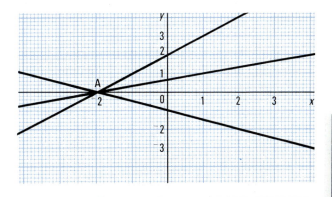

7   Find the equation of each line or curve.

a)
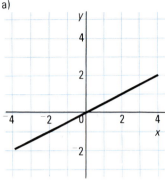

b)
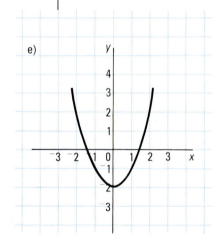

c)
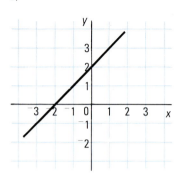

d)

e)

f)
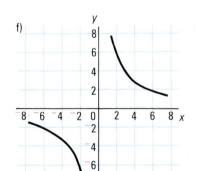

8   Draw the lines or curves which have these equations:

a) $2x - y = 3$ b) $x + y^2 = 0$ c) $y + 3x = 0$.

Lines parallel to the *x* and *y* axes

1   Line A is parallel to the *x* axis and line B is parallel
to the *y* axis. These are some of the points on line A:

($^-$10, 4) ($^-$8, 4) ($^-$7.2, 4) ($^-$1.1, 4) (0, 4)
(1.5, 4) (1.7, 4)

a) Between you list 10 points on line B.

b) Decide between you a suitable way of writing the
equation of line A and the equation of line B.

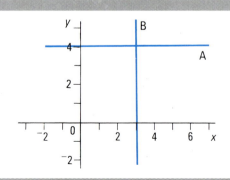

B16

The y coordinate of each point on line P is 6.
The x coordinate can be any value, for example,

$(^-17, 6)$ $(^-4.8, 6)$ $(0, 6)$ $(1.2, 6)$ $(3.8, 6)$

We can write the equation of P in words as: y is always 6, x has any value.
However, we normally write $y = 6$.
Similarly, the equation of Q is $x = 5$.

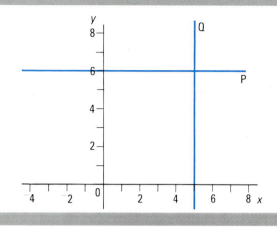

2   Write down the equations of lines A-F.

3   The lines $y = 10$, $x = ^-4$, and $y = x$ enclose a triangle. What is its area?

4   Write down the equations of four lines which enclose a *square* of area

a) 16 units²   b) 20 units².

5   What is the equation of

a) the x axis   b) the y axis?

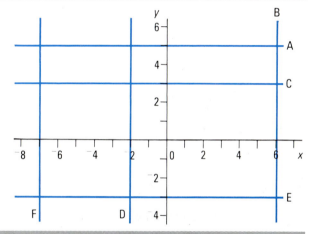

6   A, B, C and D are four vertices of an octagon which has the x axis and the y axis as lines of symmetry. The lines $y + x = 12$, $x = 8$ and $y = 8$, and five other lines, enclose the hexagon.

a) What are the coordinates of the other four vertices?

b) What are the equations of the other five lines?

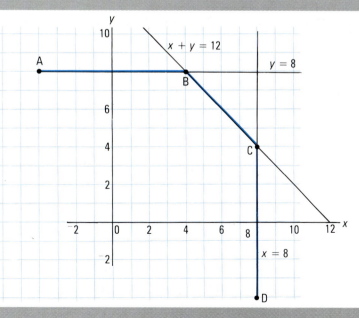

B16

7   Where do these lines   a) $x = 6$, $y = 4$ b) $2x - 5 = y$ and $x = 7$
     or curves intersect?
                              c) $y = 9$ and $xy = 9$ d) $y = 19$ and $y = x^2 + 3$ (two points).

ENRICHMENT

EXPLORATIONS

1   a) Check that the equation of line A is
        $2y = 3x + 3$.

     b) Investigate the equations of lines which
        are parallel to A. What is special about
        their equations?

     c) Investigate some more sets of parallel
        lines. Write down what is special about
        their equations.

     d) What does the equation of a line tell you
        about its slope (gradient)? Write a short
        report, with examples, to explain.

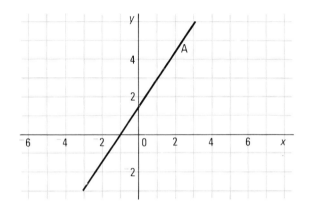

2   a) Investigate the graphs of equations like these: $y = x^2$ $y = 2x^2$ $y = 6x^2$
                                                           $y = x^2 + 1$ $y = 2x^2 + 1$ $y = 6x^2 + 1$
        Write a report about what you discover. $y = x^2 - 2$ $y = 2x^2 - 2$ $y = 6x^2 - 2$

     b) Investigate the graphs of equations like these:

        $x = y^2$ $x = 2y^2$ $x = 3y^2 + 7$

        Write a report about what you discover.

WITH A FRIEND

3   a) Investigate together graphs of equations like these:

        $x^2 + y^2 = 25$ $x^2 + y^2 = 64$

        For example, $(0, 5)\ (0,\ ^-5)\ (4, 3)\ (^-4,\ ^-3)\ (2, \sqrt{21})$ fit the equation $x^2 + y^2 = 25$.

        Write a joint report about what you discover.

     b) Investigate together graphs of equations like these:

        $x^2 - y^2 = 25$ $x^2 - y^2 = 64$

        Write a joint report about what you discover.

CORE B17

████ ACTIVITY ████

1 You will need plain paper, a pair of compasses, pencil and paper.

   a) Draw nine large triangles
      on plain paper.
      Make three of your
      triangles acute-angled
      triangles like these,

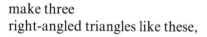

      make three
      right-angled triangles like these,

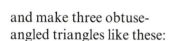

      and make three obtuse-
      angled triangles like these:

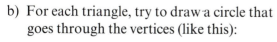

   b) For each triangle, try to draw a circle that
      goes through the vertices (like this):

      Mark the centre.

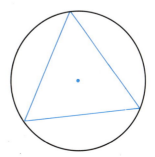

   c) Can you always draw a circle? Write
      about the position of the centres for the
      different triangles.

B17

2   Parallelogram ABCD has been drawn twice. E is where the diagonals cross.

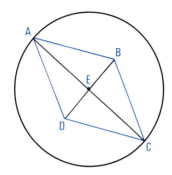
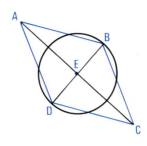

   a) With E as centre, why is it possible to draw a circle through A and C? Write one or two sentences to explain why.

   b) With E as centre, why is it possible to draw a circle through D and B?

   c) With E as centre, it is not possible to draw a circle through A, B, C and D. Why not?

   d) For some special kinds of quadrilaterals (such as squares) it *is* possible to draw a circle through A, B, C and D. What is special about these quadrilaterals? Draw diagrams to show that the circle does fit.

3   PQRS is a rectangle. The diagonals meet at T.

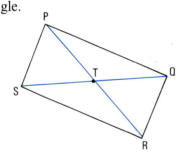
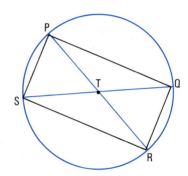

   With T as centre it is possible to draw a circle through P, Q, R and S. Why?

─────────── THINK IT THROUGH ───────────

4   MN is the diameter of the circle.

   Make a rough copy of the diagram. Draw a line through L and T so that it cuts the circle at K. What kind of figure is LMKN?

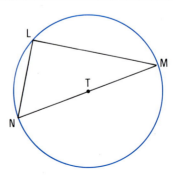

5   Draw a large circle on plain paper. Then draw a diameter and mark a point on the circumference.

Complete the triangle. Check that the marked angle measures 90°.

Mark some more points on the circumference.

Complete the triangles. Check that the marked angles measure 90°.

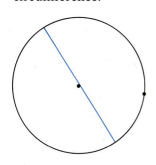
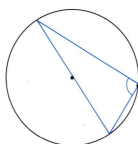
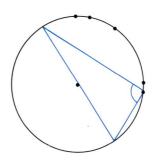
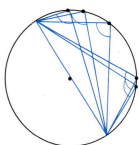

6   Copy and complete this *Take note*.

TAKE NOTE

If **AB** is any ... of a circle, and if **X** is a point on the ... of the circle, then angle AXB is a ... angle.

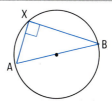

7   In these diagrams, O is the centre of the circle. Write down what you can about the size of the angle at C for each diagram i.e.
   is it ● less than 90°
        ● larger than 90°
   or ● equal to 90°?

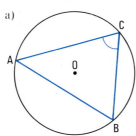

a)

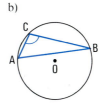

b)

c)

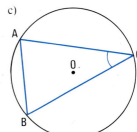

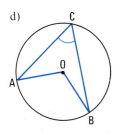

d)

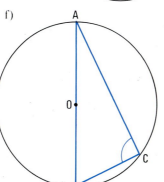

f)

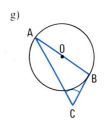

g)

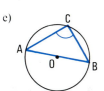

e)

━━━━━━━ EXPLORATION ━━━━━━━

8 a) Draw a circle and mark the centre. Mark any starting point inside the circle.

   b) Draw two chords through your starting point.

   c) Join the resulting points on the circumference to form a quadrilateral.

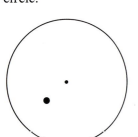
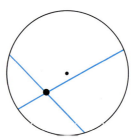
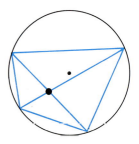

   d) Use the approach of parts a) to c) to draw some (i) kites, (ii) trapeziums, (iii) rectangles, (iv) squares.

   e) Write about the position of your starting points and the direction of the chords.

━━━━━━━ ACTIVITY ━━━━━━━

9 You will need two plain sheets of paper.

   a) (i) On plain paper draw two dots about 10 cm apart.

      (ii) Tear a corner off your other piece of paper.

      (iii) Make the two straight edges of your torn paper *touch* the two dots. Mark the position of the corner.

      (iv) Mark some more dots in the same way.

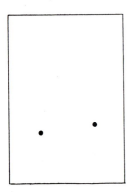
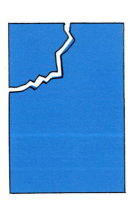
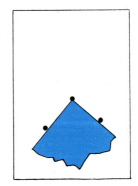
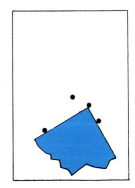

   b) When you have marked about 15 to 20 dots, join them with a smooth curve. What shape have you drawn?

   c) Try the same activity with an angle (i) less than 90° (ii) more than 90°.

      What happens this time?

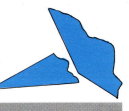

THINK IT THROUGH

10 Here is a circle.

Which of the points U to Z will form a diameter with point A? Use your corner from question 9 to find out.

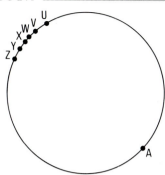

11 In this diagram, O is the centre of the circle.

This means that AB is a diameter and this means ∠ACB = 90°. If ∠CAB = 20°, what is the size of ∠CBA?

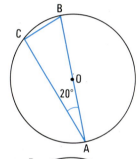

12 In this diagram, O is the centre of the circle.

∠QRO = 25°. What is the size of ∠RPQ?

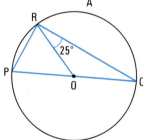

13 In this diagram, O is the centre of the circle.

Is ABCD a rectangle? How can you tell?

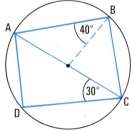

14 G and H are the centres of the circles in this diagram. How many 90° angles can you see?

Name each one.

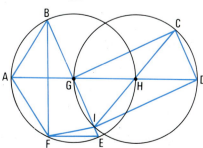

Tangents

1   a) Here is a circle and a line. (Imagine the line goes on
        forever.) The figure is folded so that the circle folds onto
        itself, *and* the line folds onto itself.

    (i)   Sketch the circle, line and fold line.
    (ii) What special point does the fold line go through?
    (iii) What is the angle between the line and the fold line?

  b) Repeat a) for
    (i) this circle and line   (ii) this circle and line.

2   a) Draw a circle with   b) Draw the radius from
        radius about 5 cm, then   the centre of the circle
        draw a *tangent* to the   to the point where the
        circle (a tangent is a   tangent touches the
        straight line that   circle.
        touches the circle at just
        one point).

  c) Measure the angle between the tangent and the radius.

3   a) Draw a circle with centre O and radius OA.

  b) Draw a straight line through A, but *not* at 90° to the radius.

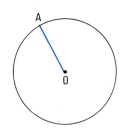

    (i) At how many points does your line meet the circle?
    (ii) Is it possible for such a line to be a tangent to the circle?

4   Copy and complete the *Take Note*.

A tangent to a circle is a straight line which
touches the circle at just ... point.

The angle between a tangent and the radius drawn
through the point of contact measures ...°.

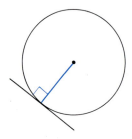

ACTIVITIES

5 a) Draw two straight lines.

   b) Draw a circle that touches both lines. (Extend the lines if necessary.)

   c) Draw some more circles that touch both lines. What do you notice about the centres?

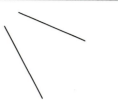

6 a) Draw three straight lines.

   b) Draw a circle that touches all three lines. (Extend the lines if necessary.)

   c) How many such circles can you draw?

7 Explain how you could find the centres in questions 5 and 6 by folding.

CHALLENGE

8 In the diagram, what is the value of $k°$? Why?

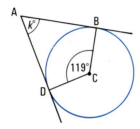

ENRICHMENT

EXPLORATION

1 On tracing paper draw some quadrilaterals which fit around a circle touching each side, and some which do not touch every side.

  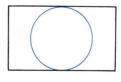

Investigate by folding what is special about the first type of quadrilateral.
Explain how you can tell, by folding, whether or not a circle can be drawn inside a quadrilateral so that it touches each side.

B18 PROBABILITY

REVIEW

- The probability that something will happen lies between 0 and 1.

Impossible   Evens   Certain

0   0.5   1

  The probability of impossible things happening (for example, you will live for a thousand years) is 0.

  The probability that something which is certain to happen will happen (for example, you will breathe in, in a few moments) is 1. The probability that a coin will turn up Heads is 0.5 or 'evens'.

■ Name your own example of an event which has a probability of happening of

     a) 1   b) 0   c) 0.5.

- If there are n equally likely outcomes, the probability of one of these happening is $\dfrac{1}{n}$.

  For example, the probability of spinning 4 on the spinner is $\dfrac{1}{6}$, because there are 6 equally likely outcomes, 1, 2, 3, 4, 5 and 6, and 4 is just one of them.

■ What is the probability that you will throw 3 on the spinner?

CORE

Using *and* and *or*

1   a) You spin a coin.
     Explain why the probability that you will spin *either* H (Heads) *or* T (Tails) is 1.

   b) You now spin two coins: a 2p and a 1p.
     Complete this list of possible outcomes: Ht, Th, . . . Explain why the probability that you spin *either* two Heads *or* two Tails *or* a Head and a Tail (on either coin) is 1.

   c) We will write P(H *or* T) = 1 to mean that the probability of getting *either* Heads *or* Tails is 1.

     (i) Explain why, when spinning a single coin, P(H *and* T) (the probability of getting H *and* T) is 0.
     (ii) What is P(H)?

   d) In b) we can write P(Hh *or* Tt *or* Ht *or* Th) = 1.

     What is   (i) P(Hh *and* Tt)
               (ii) P(Hh)
               (iii) P(Hh *or* Tt)?

B18

━━━━━━━━━━━ TAKE NOTE ━━━━━━━━━━━

- When we spin a coin we write the probability of getting H (Heads) as $P(H)$.
  $P(H) = \frac{1}{2}$, because there is one possible outcome which is acceptable out of two equally likely possibilities.

- We write $P(H \ or \ T)$ to mean the probability of getting either H or T. $P(H \ or \ T) = 1$, because it is certain we will get either H or T.

- We write $P(H \ and \ T)$ to mean the probability of getting H and T together. $P(H \ and \ T) = 0$, because it is impossible for us to get H and T together on one throw of a coin.

- When we throw a die, $P(\boxed{\,\cdot\,} \ or \ \boxed{\cdot\,\cdot}) = \frac{1}{6} + \frac{1}{6} = \frac{2}{6} \ (or \ \frac{1}{3})$

  $P(\boxed{\,\cdot\,} \ and \ \boxed{\cdot\,\cdot}) = 0.$

2   Read the *Take note*.

Explain why $P(\boxed{\,\cdot\,} \ or \ \boxed{\cdot\,\cdot}) = \frac{2}{6}$, and why $P(\boxed{\,\cdot\,} \ and \ \boxed{\cdot\,\cdot}) = 0$.

3   You spin a spinner.
The probability of spinning Red, $P(\text{Red})$, is $\frac{4}{7}$.
What is $P(not \ \text{Red})$ (that is, the probability of not spinning Red)?

━━━━━━━━━━━ TAKE NOTE ━━━━━━━━━━━

The probability of throwing $\boxed{\cdot\,\cdot}$ on a die is $\frac{1}{6}$.

The probability of NOT throwing $\boxed{\cdot\,\cdot}$ is $1 - \frac{1}{6}$, or $\frac{5}{6}$ (we can get $\boxed{\,\cdot\,}$, $\boxed{\cdot\cdot\cdot}$, $\boxed{::}$, $\boxed{\cdot\cdot\cdot\cdot}$ or $\boxed{:::}$).

4   Imagine throwing a die and spinning a coin.

a) List the possible outcomes like this: $\boxed{\,\cdot\,}$ T, $\boxed{\,\cdot\,}$ H, $\boxed{\cdot\,\cdot}$ T, $\boxed{\cdot\,\cdot}$ H, ...

b) Explain why   (i) $P(\boxed{::}) = \frac{1}{6}$

(ii) $P(T) = \frac{1}{2}$

(iii) $P(\boxed{::} \ T) = \frac{1}{12}$

(iv) $P(\boxed{::} \ T \ or \ \boxed{\cdot\cdot\cdot} \ T) = \frac{1}{6}$

(v) $P(\boxed{::} \ T \ and \ \boxed{\cdot\cdot\cdot} \ T) = 0$

5   Imagine you spin the two spinners.

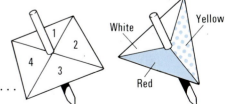

a) List all the possible outcomes, like this: ① Yellow, ① Red, ...

b) What
is: (i) $P(① \ and \ \text{Yellow})$ (iii) $P(\text{Yellow})$ (that is, Yellow and any number)
     (ii) $P(①)$ (that is, ① and any colour)   (iv) $P(① \ or \ \text{Yellow})$?

▓▓▓▓▓▓▓▓▓▓▓ TAKE NOTE ▓▓▓▓▓▓▓▓▓▓▓

You might have found 5 b) (iv) difficult to answer, because you might not have been sure if you were allowed to include ① Yellow in the acceptable outcomes.

The possible outcomes are ① Red, ① Yellow, ① White, ② Red, ② Yellow, ② White, ③ Red, ③ Yellow, ③ White, ④ Red, ④ Yellow, ④ White.

If $P(①$ or Yellow) allows ① Yellow, then $P(①$ or Yellow$) = \frac{6}{12} (\frac{1}{2})$.

If $P(①$ or Yellow) does not allow ① Yellow, then $P(①$ or Yellow$) = \frac{5}{12}$.
When we write $P(x$ or $y)$ we normally mean we allow x or y or *both*. So $P(①$ or Yellow$) = \frac{6}{12}$.

B18

7   You spin a coin and throw a die. These are all the possible outcomes:

⊡ T,   ⊡ H,   ⊡ T,   ⊡ H,

⊡ T,   ⊡ H,   ⊡ T,   ⊡ H,

⊡ T,   ⊡ H,   ⊡ T,   ⊡ H.

a) Explain why: (i) $P(⊡$ or T$)$ is $\frac{7}{12}$
                   (ii) $P(⊡$ *and* T$)$ is $\frac{1}{12}$.

b) What is: (i) $P(⊡$ or H$)$
              (ii) $P(⊡$ *and* H$)$
              (iii) $P(⊡$ T *or* ⊡ H$)$
              (iv) $P(⊡$ T *and* ⊡ H$)$?

8   a) On a spinner, $P(\text{Red}) = \frac{1}{4}$ and $P(\text{Blue}) = \frac{3}{4}$.

    What is: (i) $P(\text{Red } or \text{ Blue})$
               (ii) $P(not \text{ Red})$
               (iii) $P(not \text{ Blue})$?

    b) For another spinner $P(\text{Red}) = \frac{1}{8}$ and $P(\text{Blue}) = \frac{3}{8}$. The other colour is Green.

        (i) Sketch a possible spinner which agrees with this information.
        (ii) What is $P(\text{Green})$?
        (iii) What is $P(\text{Red } or \text{ Blue})$?
        (vi) What is $P(not \text{ Green})$?
        (v) What is $P(not \text{ Red})$?
        (vi) What is $P(\text{Red } and \text{ Green})$?

▓▓▓▓▓▓▓▓▓▓▓▓▓▓ WITH A FRIEND: EXPERIMENT CHALLENGE ▓▓▓▓▓▓▓▓▓▓▓

9   I carry out an experiment which involves two events (such as choosing a coloured bean from a bag, and then another; or spinning a coin twice; or throwing a die and spinning a spinner).

Experiment A
The probability that I succeed in the first event is $\frac{1}{4}$.
The probability that I succeed in the second event is $\frac{1}{3}$.
The probability that I succeed in both events is $\frac{1}{12}$.
Together, think of an experiment which would give this result. Describe each event carefully (that is, say what each event consists of – spinning a spinner, spinning a coin, . . . and why the probabilities are $\frac{1}{4}$, $\frac{1}{3}$, and $\frac{1}{12}$).

Experiment B
The probability that I succeed in the first event is $\frac{1}{4}$.
The probability that I succeed in the second event is $\frac{1}{3}$.
The probability that I succeed in both events is 0.
Together, think of an experiment which would give this result. Describe the experiment, and each event, carefully.

10   The diagram shows a system of tunnels.

I am equally likely to take any one turning whenever I reach a junction.

a) What is the probability that I choose route Y?

b) When I am at point Q what is the probability that I will choose the route to chamber A?

c) When I am at point P, what is the probability that I will choose the route to chamber E?

d)   (i) Check that the probability that I choose route X is $\frac{1}{2}$.
     (ii) Check that the probability that I choose the route to chamber C when I am at P is $\frac{1}{3}$.
     (iii) Check that the probability that I choose route Y and then the route to chamber E is 0.
           (You might like to compare the result in d)(iii) with your own ideas for question 9 Experiment B.)

▓▓▓▓▓▓▓▓▓▓▓▓▓▓ THINK IT THROUGH ▓▓▓▓▓▓▓▓▓▓▓

11 a) The probability of success in event A is $\frac{1}{2}$. The probability of success in event B is $\frac{1}{4}$.

    (i) Which probability is greatest:

    P(A)   P(B)   P(A *and* B)   or   P(A *or* B)?

    (ii) Which probability is smallest?

b) Success in event A means spinning Blue on spinner X.
Success in event B means spinning Blue on spinner Y.
Explain why your answers in a) are correct for spinning the two spinners.

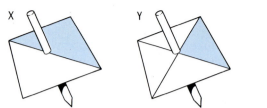

TAKE NOTE

The spinner experiment in question 11 b) has two events: Spinning X and Spinning Y. We say that the events are *independent*, because the outcome of event B does not depend upon the outcome of event A (and vice versa). For independent events, $P(A \text{ and } B)$ will always be less than $P(A)$, $P(B)$ and $P(A \text{ or } B)$ (unless $P(A)$ or $P(B)$ are 0 or 1).

The two events in the tunnel experiment in question 10 c) are *not* independent. The outcome of event B (that is, making a choice of a route) is dependent upon the outcome of event A (making the first choice of route X or Y).

---

12 Give your own example of a combined experiment which involves:

a) two independent events

b) two events which are not independent.

ENRICHMENT

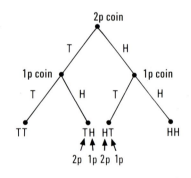

1 a) A 2p coin is spun and then a 1p coin. The experiment is repeated 20 times. The diagram represents the possible outcomes TT, TH, HT, HH.

(i) Are the two events independent or dependent events?
(ii) Explain why we would expect to get HH about 5 times.

b) Wendy uses the *tree diagram* to explain the answer to a)(ii).

For example, she uses T ⑩ to mean that out of the 20 trials, Tails is expected on the 2p 10 times.
What does the T ⑤ shown on the diagram mean?

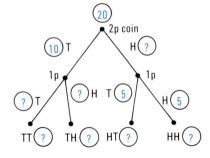

c) Copy and complete the diagram.

d) Explain why the probability of getting TT is $\frac{5}{20}$ (or $\frac{1}{4}$) (that is, $P(TT) = \frac{1}{4}$).

e) Explain why $P(TT \text{ or } HH)$ is $\frac{10}{20}$ (or $\frac{1}{2}$).

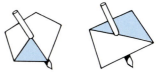

2 a) The two spinners are spun together. One possible outcome
    is White-Blue. List the other possible outcomes.

  b) About how many times would you expect White-Blue
    (White on spinner A and Blue on spinner B) in 100 trials?

  c) This how Pardip and Wendy explain their
    answers to a) and b):

Pardip

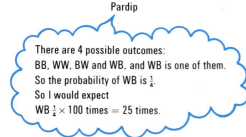

There are 4 possible outcomes:
BB, WW, BW and WB, and WB is one of them.
So the probability of WB is $\frac{1}{4}$.
So I would expect
WB $\frac{1}{4}\times 100$ times = 25 times.

Wendy

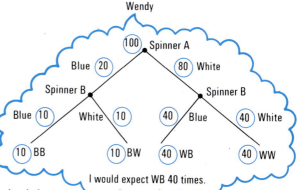

I would expect WB 40 times.

    Explain why Pardip is wrong and why Wendy is right.

  d) Explain why $P(\text{BW})$ is $\frac{1}{10}$.

  e) What is (i) $P(\text{WW})$ (ii) $P(\text{BW } or \text{ WB})$?

━━━━━━━━━ TAKE NOTE ━━━━━━━━━

Before making probability calculations check carefully to see if outcomes are equally likely or not.

3 You spin all three spinners together 600
  times.

  a) Copy and complete the diagram to show
    about how many times you should expect
    each possible outcome.

  b) About how many times
    in the 600 trials should
    you expect to score

    (i) White on exactly
      two spinners

    (ii) Red on exactly two
       spinners

    (iii) White on more
        than one spinner

    (iv) no Reds?

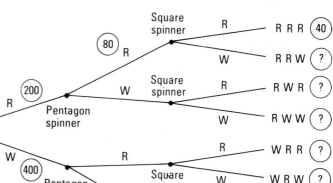

  c) Explain why the probability of getting White on all three
    spinners *or* Red on all three spinners is $\frac{160}{600}$ (or $\frac{4}{15}$).

B18

4   The probability that you
will score Red on spinner A is $\frac{1}{3}$,
and on spinner B is $\frac{3}{8}$.

A   B

a) The possible outcomes are RW, WR, RR
and WW. Explain why the probability
that you will score WW is *not* $\frac{1}{4}$.

b) How many times would you expect to score WW in 240 experiments? Explain your result.

c) Imagine that the spinners are spun 120
times. Copy and complete the tree
diagram.

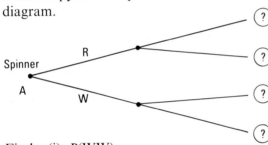

d) Find   (i) *P*(WW)
(ii) *P*(RW *or* WR)
(iii) the probability of getting Red on both spinners
(iv) the probability of getting Red on at least one spinner.

━━━━━━━━━ WITH A FRIEND ━━━━━━━━━

5   a) Discuss the connection between these calculations, and the situation in question 4. Each of
you write down what you decide. Try to explain why the connections exist.

(i) $\frac{1}{3}$ of $120 = 40$
(ii) $\frac{3}{8}$ of $\frac{1}{3}$ of $120 = 15$
(iii) $\frac{2}{3}$ of $120 = 80$
(iv) $\frac{3}{8}$ of $\frac{2}{3}$ of $120 = 30$

b) Write down two more sets of calculations which relate to the situation in question 4.

━━━━━━━━━ CHALLENGE ━━━━━━━━━

6   The probability that you score Red on a spinner is $\frac{3}{10}$. The probability that you score Red on a
second spinner is $\frac{3}{8}$. About how many times in 800 experiments would you expect to score
Red-Red?

7   The probability that it will rain during the Wimbledon tennis tournament is $\frac{3}{8}$. The probability that it will rain during the Old Trafford Test is $\frac{2}{5}$.

a) Imagine a 40-year period. In about how many of these years would you expect it to rain

  (i) during the Wimbledon tennis tournament
  (ii) during the Old Trafford Test
  (iii) on both occasions?

b) Copy and complete the tree diagram.

c) Use your results from the tree diagram to explain why the probability that it will *not* rain at either Wimbledon or Old Trafford is $\frac{15}{40}$ (or $\frac{3}{8}$).

d) What is the probability that

  (i) it will rain at both venues
  (ii) it will rain at just one venue
  (iii) it will rain in at least one venue?

e) About how many times during each century would you expect it to rain both during the Wimbledon tennis tournament and at the Old Trafford Test?

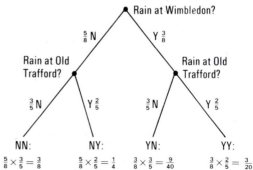

TAKE NOTE

When an experiment consists of two independent events A and B, we find the probability of pairs of outcomes by multiplying the probabilities for each outcome. We multiply 'down the branches of the tree diagram':

P(Rain at Wimbledon *and* Rain at Old Trafford)

$= \frac{3}{8} \times \frac{2}{5} = \frac{3}{20}$

P(Rain at Wimbledon *and* No rain at Old Trafford)

$= \frac{3}{8} \times \frac{3}{5} = \frac{9}{40}.$

Rain at Wimbledon?

$\frac{5}{8}$ N   Y $\frac{3}{8}$

Rain at Old Trafford?   Rain at Old Trafford?

$\frac{3}{5}$ N   Y $\frac{2}{5}$ $\frac{3}{5}$ N   Y $\frac{2}{5}$

NN: NY: YN: YY:

$\frac{5}{8} \times \frac{3}{5} = \frac{3}{8}$ $\frac{5}{8} \times \frac{2}{5} = \frac{1}{4}$ $\frac{3}{8} \times \frac{3}{5} = \frac{9}{40}$ $\frac{3}{8} \times \frac{2}{5} = \frac{3}{20}$

■■■■■■■■■■■■■■ THINK IT THROUGH ■■■■■■■■■■■■■■

8   Read the *Take note.*

a) Write your own *Take note* to explain how we find probabilities such as $P(YY \text{ or } NN)$, $P(YN \text{ or } NY)$ from a tree diagram. Use the 'Rain at Wimbledon and Old Trafford' tree diagram to illustrate your *Take note*.

b) In a motor car factory, the probability that a new car has faulty electrics is $\frac{3}{50}$. The probability that it has faulty mechanical parts is $\frac{1}{20}$.

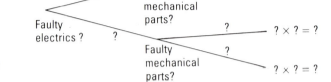

(i)   Copy and complete the tree diagram.

(ii)   Use the tree diagram to find the probability that a car you buy from the factory has

A   faulty electrics *and* faulty mechanical parts

B   *either* faulty electrics *or* faulty mechanical parts (but not both)

C   neither mechanical nor electrical faults.

CORE

=== ACTIVITY ===

1 You need plain paper.
The two ants travel around the square (in a clockwise direction) at the same speed. One starts from A and the other starts from B. P is the point midway between them at any time.

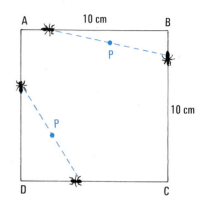

a) In your head, imagine what kind of shape P traces out as the ants travel around the square.

b) Check what you decide in a) by drawing.

c) Investigate the shape which P traces out for different starting points for the ants (with different distances between them). Write a report about what you discover.

2 a) P and Q are two tigers. You are at X, an equal distance from each. You have to walk between them, and you decide that you will keep your distance from each the same as you walk.

Make a sketch of P, Q and X, and show on your sketch the path which you must take.

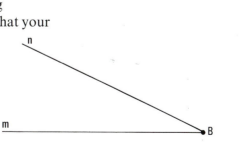

b) m and n are two walls. You are standing at B, between them. You must walk so that your distance from each wall is always equal. Sketch the path which you must take.

3 a) A moth flies so that its distance from the light is always 80 cm. Describe the set of positions which the moth can take up.

b) Another moth flies so that its distance from the light is always 80 cm or less. Describe the set of positions which this moth can take up.

████████████ TAKE NOTE ████████████

The set of positions which a point can occupy is called the locus of the point. The locus of a point can be a line (as in questions 1, 2 and 3) or a surface or solid (as in question 4). (The plural of locus is *loci* or locuses.)

██

4 a) Draw a straight line **AB**. Describe the *locus* of a point which can move so that it is always 2 cm from the line. (Make a sketch to help you to explain.)

b) Describe the locus when the point can move in 3D space rather than on the flat surface of the paper.

████████████ ACTIVITY ████████████

5 You will need some card or stiff paper. Mark two points X and Y about 3 cm apart. Cut a triangle from card or stiff paper. Move the triangle between X and Y so that PR always touches X and PQ always touches Y. Sketch the loci of P, Q and R.

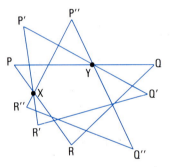

██

6 a) A 10p coin rolls around the outside of a can. What is the locus of the centre of the 10p?

b) Sketch the locus of a point on the circumference of the 10p.

7 The concrete block is too heavy to lift, so it is rolled along the ground. The cross section is a square.

a) Sketch the locus of T, the centre of the end of the block.

b) Sketch the locus of P, the centre of one of the rectangular faces of the block.

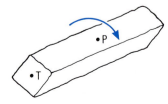

8 a) This is a clockwork train going round a circular track in a lift. What is the locus of a point on the front of the train as the lift ascends? Make a sketch to help you to explain.

   b) Sketch the locus of the tip of a propellor as an aeroplane flies horizontally.

9 Sometimes we can describe a locus with an equation. For example, the locus of a point which moves in a straight line through C, A and B is $x = y$.

Describe the locus of each of these points by an equation.

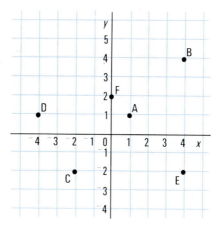

   a) The point which moves in a straight line through D and A.

   b) The point which moves in a straight line through B and E.

   c) The point which moves in a straight line through F, A and E.

===== THINK IT THROUGH =====

10 A point moves so that its distance from a horizontal line is y and its distance from a vertical line is x, and so that

   $y^2 = 4x$.

   Draw a grid and sketch the locus of the point.

B19

ENRICHMENT

━━━━━━━ ACTIVITY ━━━━━━━

1   You need a sheet of thin card, scissors and a large sheet of
    plain paper (or several sheets of A4 plain paper).
    Cut out the circular disc shown here from thin card. Punch
    holes A (at the centre), B, C and D (each further away from
    the centre than the other). Mark E somewhere on the rim.

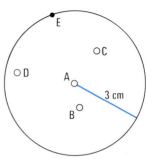

a) *Imagine* the disc rolling
    along a straight line.
    Draw 5 straight lines,
    one above the other.

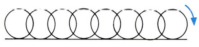

Predict the loci of A, B,
C, D and E as the disc
rolls along the line.
Make a sketch to show
each prediction. (You
don't have to make
these sketches full size.)

Locus of A ————————————

Locus of B ————————————

Locus of C ————————————

Locus of D ————————————

Locus of E ————————————

b) Now experiment using the disc. Check if your predictions in a) are correct.

2   a) A point P moves so that its distance from the line m and its
        distance from the point A are always equal. Sketch the
        locus of the point. What special name is given to this locus?

    b) A point P moves so that the sum of its distances from two
        fixed points A and B is constant (for example,
        PA + PB = 10). Sketch the locus
        of the point. What special
        name is given to this locus?

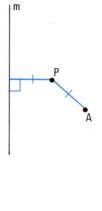

B20 FORMULAS AND GENERALISATIONS

REVIEW

Letters can be used ● to stand for unknown numbers.   For example, $5x + 1 = 8$
$$x = ?$$

● to give instructions.   For example,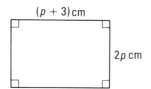
(p + 3) cm
2p cm

● to write rules.   For example,

The number of blue dots in
each collection (p) is two more
than the number of black dots (t).

$$p = t + 2$$

CONSOLIDATION

1 Solve each equation:

   a) $5x + 1 = 6$ b) $3x - 6 = 0$.

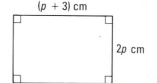
(p + 3) cm
2p cm

2 For what value of p do these instructions produce a square?

3 This ripple of rectangles was obtained
   from a set of instructions like those in
   question 2. Draw a rectangle and write in
   the instructions. Choose your own letter.

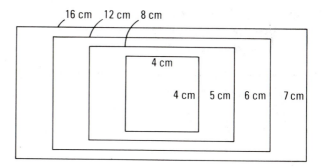
16 cm   12 cm   8 cm
4 cm
4 cm   5 cm   6 cm   7 cm

B20

CHALLENGE

4 Find the rule connecting the numbers of black dots and blue dots. Write the rule:

   a) in words

   b) using letters
       (choose your own letters).

1

3n   2

There is a connection between the collection of dot patterns (there is a never-ending line of them) and the expression $3n-2$. Decide between you what it is. Write down what you decide.

320

2   These patterns of crosses are arranged in order.

a) How many crosses does the twentieth pattern have?

b) Use the expression $2n-1$. Choose n to be 1, 2, 3, 4, . . . What do your results tell you about the pattern of crosses?

c) One of the patterns has 131 crosses. What position is the pattern in the sequence? What value of n gives this number of crosses?

3   a) Use the expression $3n-1$. Choose n to be 1, 2, 3, 4 and 5. Represent your results as a sequence of dot patterns.

b) One of the patterns in this sequence has 56 dots. What position is the pattern in the sequence?

4   a) Find an expression using n, which will produce a sequence of dot patterns of which the first four are:

b) One of the patterns has 331 dots. What position is it in the sequence?

Expressions like $3n + 1$ can be used to produce sequences of numbers by successively giving n the values 1, 2, 3, . . .

4	7	10	13	. . .
↑	↑	↑		
$n = 1$	$n = 2$	$n = 3$		

The fourth number is $(3 \times 4) + 1 = 13$.
We say that the mth number is $3m + 1$; the kth number is $3k + 1$; the nth number is $3n + 1$.

5   How many dots does (i) the 50th pattern (ii) the nth pattern, have in each of these sequences:

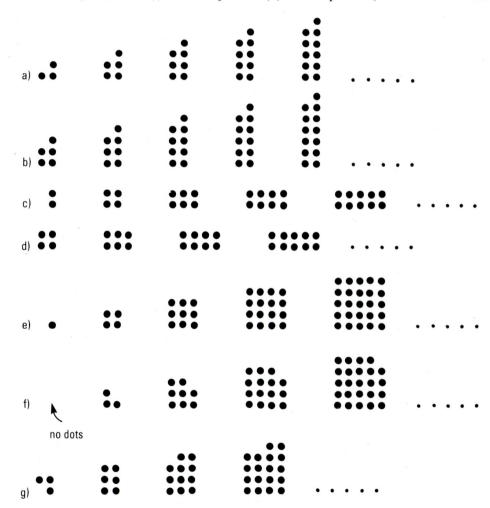

In this set of arrays, the number of dots (N) in the nth array is $2n + 1$.
We can write the formula: $N = 2n + 1$.

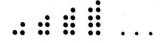

B20

6   The fifth number in this sequence of numbers is 9:

$1, 3, 5, 7, 9, 11, \ldots$

a) Check that the 20th number is $(2 \times 20) - 1 = 39$.

b) What is the 90th number?

c) Write down an expression for the nth number.

d) The nth number in the sequence is N. Write a formula connecting N and n.

7   Call the nth number in each of these sequences 'N'.
Find   a) the 90th number   b) a formula connecting N and n.

(i) $3, 5, 7, 9, 11, \ldots$ (v) $0, 3, 8, 15, 24, \ldots$
(ii) $2, 4, 6, 8, 10, \ldots$ (vi) $2, 5, 8, 11, 14, \ldots$
(iii) $4, 6, 8, 10, 12, \ldots$ (vii) $1, 4, 7, 10, 13, \ldots$
(iv) $1, 4, 9, 16, 25, \ldots$ (viii) $3, 7, 11, 15, 19, \ldots$

8   The nth number in a sequence is $n - 30$.

a) Check that the first number is $^-29$. What is the 2nd number?

b) What is   (i) the 29th number   (ii) the 30th number   (iii) the 31st number?

B20

9   Write down the first four numbers in the sequence whose nth number is:

a) $2 - n$ b) $2(n-1)$ c) $n^2 - 3$ d) $3 - n^2$ e) $\dfrac{2}{n}$.

━━━━━━━━━━ EXPLORATION ━━━━━━━━━━

10 Investigate some sequences
of your own. Plot them on
a grid like this. Find
different 'types' of
sequences which give
different 'types' of graphs.
Write a report to explain
what you discover.

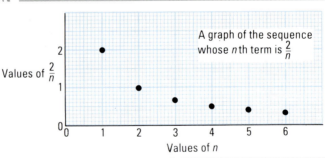

A graph of the sequence whose nth term is $\frac{2}{n}$

Values of $\frac{2}{n}$

Values of n

Finding the terms of a sequence

1   Find the next term in each of these sequences.

(i) $1, 4, 7, 10, 13, \ldots$
(ii) $2, 4, 8, 16, 32, \ldots$
(iii) $4, 3, 2\frac{1}{2}, 2\frac{1}{4}, 2\frac{1}{8}, \ldots$
(iv) $1, 1, 2, 3, 5, 8, \ldots$

2   These are the sequences in question 1.

    (i) 1   4   7   10   13 . . .

          3   3   3   3 . . .

    (ii) 2   4   8   16   32 . . .

           2   4   8   10 . . .

    (iii) 4   3 $2\frac{1}{2}$ $2\frac{1}{4}$ $2\frac{1}{8}$ . . .

           1   $\frac{1}{2}$ $\frac{1}{4}$ $\frac{1}{8}$ . . .

    (iv) 1 1 2 3 5 8 . . .

           0 1 1 2   3 . . .

a) What does each row of smaller numbers represent?

b) The smaller numbers will help you to find the next term in each sequence. Explain how, and use them to check your results in question 1.

c) Find the next term in each of these sequences. (Writing the difference between successive terms below the sequence will help.)

    (i) 2, 7, 17, 37, 77, . . .
    (ii) 2, 5, 14, 41, . . .
    (iii) 2, 3, 6, 11, 18, 27, . . .
    (iv) 1, 2, 6, 15, 31, 56, . . .

▬▬▬▬▬ CHALLENGE ▬▬▬▬▬

3   Find the 20th term in each sequence:

a) 2, 5, 10, 17, 26, 37, 50, . . .

b) 0, 2, 6, 12, 20, 30, . . .

Writing generalisations with letters

1   a) How many dots and lines will there be in the 10th drawing?

    •   1 dot   0 lines

        2 dots   1 line

        3 dots   3 lines

        4 dots   6 lines

b) Which of these formulas gives the correct relationship between the number of lines (L) and the number of dots (n)?

    (i) $L = n - 1$ (ii) $L = \dfrac{n}{2}$ (iii) $L = n^2$ (iv) $L = n^2 + 2$ (v) $L = \dfrac{1}{2}n(n-1)$

B20

2   a) How many dots are there in the 10th array?

b) Which of these formulas gives the correct relationship beween the number of dots (N) in an array and the position (n) of the array in the sequence?

(i) $N = n$ (ii) $N = \dfrac{3n}{2}$ (iii) $N = \frac{1}{2}n(n-1)$ (iv) $N = \frac{1}{2}n(n+1)$

CHALLENGE

3   Find a formula connecting the number of dots (N) in each array to its position (n) in the sequence.

a)

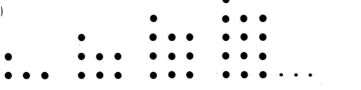

b)

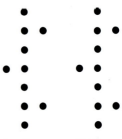

c)

4   The drawing shows some polygons divided into triangles by lines from one vertex.

a) Use n for the number of sides of any polygon. Explain why the sum of the interior angles of the polygon is $180 \times (n-2)$ degrees.

b) What is the sum of the interior angles of a 20-sided polygon?

5   a) Find a connection between number of diamonds (d) and number of dots (t) in each pattern. Write it in two different ways like this: $t = \ldots$, and $d = \ldots$

 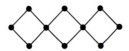 

b) When all the patterns are arranged in order the *n*th pattern will have *n* diamonds. Find an expression, using *n*, for the number of dots it has.

6 a) How many cubes are needed to build the tower?

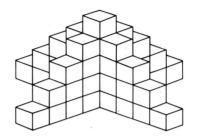

   b) How many cubes are needed to build a similar tower 12 cubes high?

   c) Find an expression for the number of cubes which are needed for a tower *n* cubes high.

7 Inside this 4×4 square grid there are sixteen 1×1 squares
nine 2×2 squares
and four 3×3 squares.

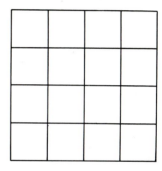

Altogether, then, there are thirty squares (counting the 4×4 square). Investigate how many squares there are for other square grids (for example, 2×2 grid, 5×5 grid, etc.).

For an $n \times n$ grid, how many

1×1, 2×2, 3×3, $(n-1) \times (n-1)$, $n \times n$

squares are there? What is the total number of squares?

B20

ENRICHMENT

1 You need squared paper.
Mark out different-sized rectangles on squared paper and draw a diagonal. Find the number of squares (*n*) which the diagonal crosses. Make a table of your results.

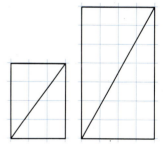

Find relationships connecting the rectangle sizes and the number of squares crossed for different groups of rectangles (for example, rectangles of width 1 square, width 2 squares, width 3 squares, . . .) Is there a general rule for all rectangles?

2 The '$3 \times 3 \times 3$' cube is made out of 1 cm blocks. The outside of the cube is painted blue.

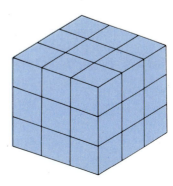

How many blocks have 3 faces painted?
2 faces painted?
1 face painted?
0 faces painted?

Investigate the same questions for different sizes of cube. Write down the results for an '$n \times n \times n$' cube.

REVIEW

- The plane has been *rotated* through 90° clockwise about 0. The point A(2, 1) has been mapped onto A′(1, ⁻2). We say the image of A(2, 1) is A′(1, ⁻2).

■ What is the image of B(2, 2) under the rotation?

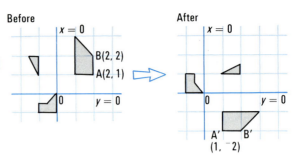

- The plane has been *reflected* in the line $x = 0$. The image of A(2, 1) is A′(⁻2, 1).

■ What is the image of B(2, 2)?

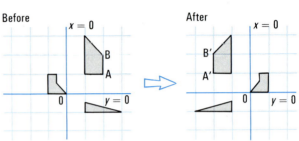

- The plane has been *translated* using the translation $\binom{2}{1}$. The image of A(2, 1) is A′(4, 2).

■ What is the image of B(2, 2)?

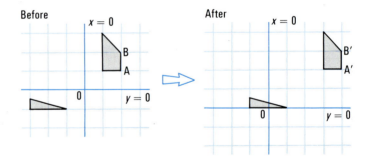

- The plane has been *enlarged*, scale factor × 2, centre (0, 0). The image of A(2, 1) is A′(4, 2).

■ What is the image of B(2, 2)?

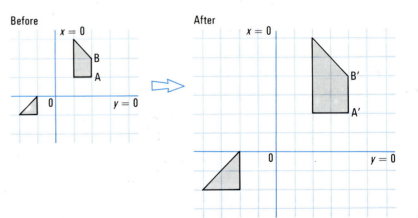

CORE

1   Write down the coordinates of the vertices of triangle A′B′C′ for each transformation. The new position of one vertex is given to help you in a) to c).

a)

  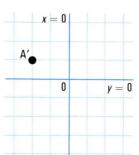

Rotation 90° anticlockwise, centre (0, 0)

b)

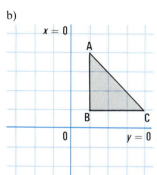 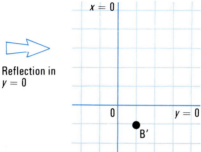

Reflection in y = 0

c)

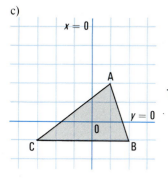 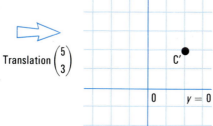

Translation $\binom{5}{3}$

d)

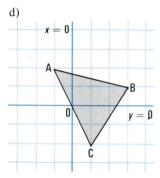 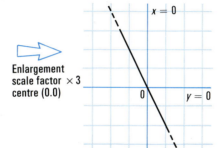

Enlargement scale factor × 3 centre (0.0)

B21

CHALLENGE

2   Each diagram represents a transformation of a figure.
    Describe each transformation accurately. For example, 'a half turn, centre (1, 0)'...

a)
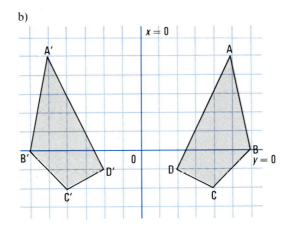

b)

c)
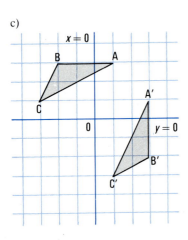

d)
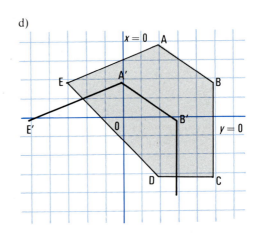

e)
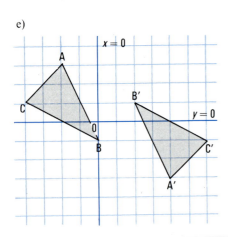

f)
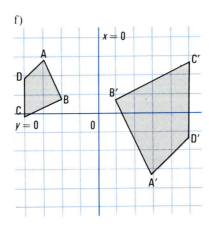

B21

3   You need squared paper.
In each of these situations you are given a set of points and their images after a transformation.
Use squared paper to help you to describe each transformation accurately.

Transformation A
$(1, 1) \longrightarrow (1, ^-1)$
$(0, 4) \longrightarrow (4, 0)$
$(^-2, 3) \longrightarrow (3, 2)$

> The image of (1,1) is (1, ̄1).
> We say (1,1) is mapped onto
> (1, ̄1).

Transformation B
$(1, 1) \longrightarrow (1, 1)$
$(^-3, 4) \longrightarrow (5, 4)$
$(0, ^-7) \longrightarrow (2, ^-7)$

Transformation C
$(4, 7) \longrightarrow (6, 10)$
$(0, ^-1) \longrightarrow (2, 2)$
$(^-1, ^-2) \longrightarrow (1, 1)$

Transformation D
$(0, 4) \longrightarrow (2, 0)$
$(4, 4) \longrightarrow (^-2, 0)$
$(^-6, 2) \longrightarrow (8, 2)$

Transformation E
$(2, 1) \longrightarrow (^-1, ^-2)$
$(5, 0) \longrightarrow (0, ^-5)$
$(^-4, 4) \longrightarrow (^-4, 4)$

Transformation F
$(2, 2) \longrightarrow (8, 2)$
$(2, 0) \longrightarrow (8, ^-6)$
$(^-1, 1) \longrightarrow (^-4, ^-2)$

WITH A FRIEND: THE TRANSFORMATIONS GAME

4   You both need squared paper for rough working.
There are five stages to the transformation game.

STAGE 1   Reflection
STAGE 2   Rotation
STAGE 3   Translation
STAGE 4   Enlargement
STAGE 5   Your choice

At STAGE 1 you both choose a reflection transformation (for example, reflection in $x+y = 4$)
and write it down, secretly. In turn, ask each other for the image of a point under the
transformation (for example, 'What is the image of (0, 0) under your transformation?'). The aim
is to identify your friend's transformation after as few questions as possible.

Scoring   Correct identification after 1 question   - 5 points   3 questions - 3 points
                                          2 questions   - 4 points   4 questions - 1 point.

STAGES 2-4 follow the same rules for rotation, translation and enlargement transformations,
and at STAGE 5 you can choose any kind of transformation you wish - and you don't have to tell
your friend which type you have chosen. After STAGE 5, the person with more points wins.

B21

ENRICHMENT

1 a) The diagram represents the result of a
       double transformation of the plane.

       In the first transformation triangle
       ABC becomes triangle A′B′C′. In the
       second transformation triangle A′B′C′
       becomes triangle A″B″C″.
       What kinds of transformations were applied
       to the plane? Describe each one accurately.

   b) What *single* transformation would have
       produced the same final result?

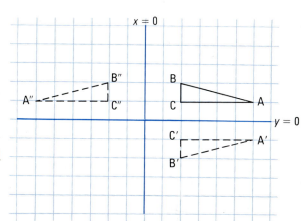

2 Draw a diagram which shows that a reflection in the vertical axis, followed by a reflection in the horizontal axis, is equivalent to a half turn, centre $(0, 0)$. Choose your own figure in the plane to transform.

CHALLENGE

3 X means a reflection in the horizontal axis.
Y means a reflection in the vertical axis.
H means a half turn about $(0, 0)$.
R_1 means a quarter turn clockwise about $(0, 0)$.
R_2 means a quarter turn anticlockwise about $(0, 0)$.
P means a reflection in the line through $(0, 0)$ and $(1, 1)$ (that is, $x = y$).
Q means a reflection in the line through $(0, 0)$ and $(1, ^-1)$ (that is, $x + y = 0$).
I means that every point in the plane is invariant (remains in the same position).

Investigate the results of combinations of these transformations.

For example, what single transformation is equivalent to X followed by H? Write your results in a table, like this.

		Second transformation						
		X	Y	H	R_1	R_2	P	Q
	X	I						
	Y							
First transformation	H							
	R_1							
	R_2							

B21